博士后文库
中国博士后科学基金资助出版

U0140735

Monoidal Invariants on Representation Categories of Hopf Algebras

(Hopf 代数表示范畴中的 Monoidal 不变量)

Zhihua Wang (王志华)

Science Press
Beijing

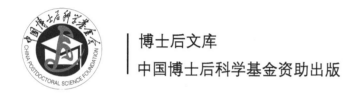

Responsible Editor: Qingjia Hu

Published by Science Press
16 Donghuangchenggen North Street
Beijing 100717, P. R. China

Printed in Beijing

ISBN 978-7-03-078146-8

作 者 简 介

王志华，男，1980 年 11 月出生，江苏东台人，泰州学院教授. 2014 年取得比利时 Hasselt 大学博士学位，2015 年取得扬州大学博士学位，2017—2019 年进入南京大学博士后流动站从事博士后工作，2019 年入选江苏高校“青蓝工程”中青年学术带头人培养对象. 工作以来在国内外期刊上发表学术论文四十多篇.

《博士后文库》序言

1985 年, 在李政道先生的倡议和邓小平同志的亲自关怀下, 我国建立了博士后制度, 同时设立了博士后科学基金. 30 多年来, 在党和国家的高度重视下, 在社会各方面的关心和支持下, 博士后制度为我国培养了一大批青年高层次创新人才. 在这一过程中, 博士后科学基金发挥了不可替代的独特作用.

博士后科学基金是中国特色博士后制度的重要组成部分, 专门用于资助博士后研究人员开展创新探索. 博士后科学基金的资助, 对正处于独立科研生涯起步阶段的博士后研究人员来说, 适逢其时, 有利于培养他们独立的科研人格、在选题方面的竞争意识以及负责的精神, 是他们独立从事科研工作的“第一桶金”. 尽管博士后科学基金资助金额不大, 但对博士后青年创新人才的培养和激励作用不可估量. 四两拨千斤, 博士后科学基金有效地推动了博士后研究人员迅速成长为高水平的研究人才, “小基金发挥了大作用”.

在博士后科学基金的资助下, 博士后研究人员的优秀学术成果不断涌现. 2013 年, 为提高博士后科学基金的资助效益, 中国博士后科学基金会联合科学出版社开展了博士后优秀学术专著出版资助工作, 通过专家评审遴选出优秀的博士后学术著作, 收入《博士后文库》, 由博士后科学基金资助、科学出版社出版. 我们希望, 借此打造专属于博士后学术创新的旗舰图书品牌, 激励博士后研究人员潜心科研, 扎实治学, 提升博士后优秀学术成果的社会影响力.

2015 年, 国务院办公厅印发了《关于改革完善博士后制度的意见》(国办发〔2015〕87 号), 将“实施自然科学、人文社会科学优秀博士后论著出版支持计划”作为“十三五”期间博士后工作的重要内容和提升博士后研究人员培养质量的重要手段, 这更加凸显了出版资助工作的意义. 我相信, 我们提供的这个出版资助平台将对博士后研究人员激发创新智慧、凝聚创新力量发挥独特的作用, 促使博士后研究人员的创新成果更好地服务于创新驱动发展战略和创新型国家的建设.

祝愿广大博士后研究人员在博士后科学基金的资助下早日成长为栋梁之才, 为实现中华民族伟大复兴的中国梦做出更大的贡献.

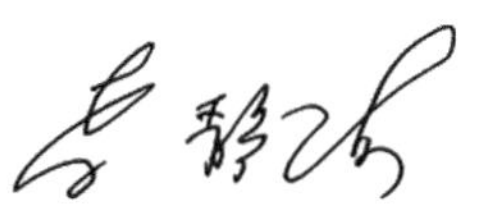

中国博士后科学基金会理事长

Preface

For any finite dimensional Hopf algebra H, the representation category H-mod consisting of all finite dimensional left H-modules is a monoidal category whose monoidal structure stems from the comultiplication of H. It can be seen from [63, Theorem 2.2] that the representation categories of two finite dimensional Hopf algebras H and H' are equivalent as monoidal categories if and only if there exists a twist J on H such that H' and the twisted Hopf algebra H^J are isomorphic as bialgebras. However, to determine the existence or nonexistence of such twists and isomorphisms is still highly non-trivial. This suggests a need of invariants for the monoidal structure of the representation categories of finite dimensional Hopf algebras.

In this book, we introduce some invariants of the representation category H-mod for a finite dimensional Hopf algebra H. These invariants include the Green rings, the Casimir numbers, the higher Frobenius-Schur indicators, the Grothendieck algebras and certain polynomial invariants. These are monoidal invariants which can be used to distinguish representation categories of Hopf algebras. Now let us introduce these invariants in more details.

In the study of the monoidal structure of H-mod, one has to consider the decompositions of tensor products of H-modules. However, in general, very little is known about how to decompose tensor products of two indecomposable H-modules into a direct sum of indecomposable H-modules. One method of addressing this problem is to consider the Green ring $r(H)$ which gives information on the decompositions of tensor products of H-modules.

The Green rings of representation categories of groups and Hopf algebras have attracted much attention. Firstly, Green [33, 34], Benson and Carlson, etc., considered the Green rings of modular representations of finite groups. Later, Witherspoon [93] studied the Green ring of the quantum double of a finite group. Wakui [82] computed the Green rings of all non-semisimple Hopf algebras of dimension 8 in terms of generators and relations. Cibils [18] determined all graded Hopf algebras on a cycle path coalgebra (which are just equal to the generalized Taft algebras (see [70])), and considered the tensor product decompositions of indecomposable representations (see [36]). Moreover, Cibils computed the Green ring of the Sweedler 4-dimensional Hopf algebra in terms of generators and relations. For Green rings of quantum groups, we refer to the work by Domokos and Lenagan [24].

Chen, Van Oystaeyen and Zhang [15] described the Green rings of Taft algebras in terms of generators and relations. The Green rings of generalized Taft algebras were later studied in [49]. The Green rings of finite dimensional pointed Hopf algebras of rank one were investigated in [86,87], while the Green rings of pointed Hopf algebras of rank two were studied in [14, 47]. For Hopf algebras of infinite representation type, their Green rings are usually not finitely generated but some of them are still computable. For example, the Green ring of the Drinfeld double of Sweedler Hopf algebra [13] and the Green rings of Drinfeld doubles of Taft algebras [79] are not finitely generated but they can be described explicitly. For more examples of Green rings of Hopf algebras and tensor categories, we refer the readers to [14, 16, 40, 41, 77, 78, 96, 97].

After Green [34] first showed that the Green ring has no nonzero nilpotent elements for any cyclic p-group over a field of characteristic p, many subsequent works have centered on the nilpotency problem, that is, whether or not the Green ring possesses nonzero nilpotent elements. The nilpotency problem has been completely solved for the Green ring of a finite group. It was shown that when the base field is of characteristic p, the Green ring of a finite group G contains nonzero nilpotent elements unless the Sylow p-subgroups of G are cyclic or elementary abelian 2-groups (see [7, 34, 98]). For the Green ring of a Hopf algebra, if H is a finite dimensional pointed Hopf algebra of rank one, then all nilpotent elements of the Green ring of H form a principal ideal, which is nothing but the Jacobson radical of the Green ring (see [86, Theorem 5.4] and [87, Theorem 6.3]). The proofs given for the above results were heavily computational, and neither explained the properties of nilpotent elements, nor indicated a criterion for detecting them.

As we shall see, the Green ring $r(H)$ is a Frobenius algebra over $\mathbb{Z}$ if the Hopf algebra H is of finite representation type. In this case, using the Casimir operator of $r(H)$, one is able to define a numerical invariant of H-mod, called the Casimir number of H. This number is a non-negative integer which can be used to determine when the Green ring $r(H)$ or the Green algebra $r(H) \otimes_{\mathbb{Z}} K$ over a field K is semisimple (with zero Jacobson radical). If $r(H)$ is a commutative ring with finite rank as a free abelian group, then the Jacobson radical of $r(H)$ is the set of all nilpotent elements of $r(H)$. In this case, the Casimir number can be used to detect whether the Green ring $r(H)$ has nonzero nilpotent elements.

A fusion category $\mathcal{C}$ over a field $\Bbbk$ is called non-degenerate if the global dimension $\dim(\mathcal{C})$ of $\mathcal{C}$ is not zero in $\Bbbk$. Since $\dim(\mathcal{C})$ is automatically not zero in a field $\Bbbk$ of characteristic zero, this notation is only considered in a field $\Bbbk$ of positive characteristic. A crucial property of non-degenerate fusion categories is that they can be lifted to the case of characteristic zero (see e.g., [30, Section 9]). It is interesting to know when a fusion category over a field of positive characteristic is non-degenerate. Ostrik stated that a spherical fusion category $\mathcal{C}$ over a field $\Bbbk$ is non-degenerated, if the Grothendieck algebra $\mathrm{Gr}(\mathcal{C}) \otimes_{\mathbb{Z}} \Bbbk$ is semisimple (see [68, Proposition 2.9]). It has been proved by Shimizu that a pivotal fusion category $\mathcal{C}$ over an algebraically closed field $\Bbbk$ is non-degenerate if and only if its Grothendieck algebra $\mathrm{Gr}(\mathcal{C}) \otimes_{\mathbb{Z}} \Bbbk$ is semisimple (see [73, Theorem 6.5]).

The notation of the Casimir number can be defined over a fusion category $\mathcal{C}$. The semisimplicity criterion for Green algebra $r(H)\otimes_{\mathbb{Z}} K$ also works for the Grothendieck algebra of $\mathcal{C}$. This criterion together with Shimizu's work [73, Theorem 6.5] gives a numerical criterion for a pivotal fusion category to be non-degenerate. That is, a pivotal fusion category $\mathcal{C}$ over a field $\Bbbk$ is non-degenerate if and only if the Casimir number of $\mathcal{C}$ is not zero in $\Bbbk$.

The Frobenius-Schur exponent of a spherical fusion category $\mathcal{C}$ has been defined in [61, Definition 5.1] as a minimal positive integer satisfying certain properties. In the case that the ground field is the field $\mathbb{C}$ of complex numbers, the Cauchy theorem for spherical fusion categories asserts that the prime ideals dividing the global dimension $\dim(\mathcal{C})$ of $\mathcal{C}$ and those dividing the Frobenius-Schur exponent of $\mathcal{C}$ are the same in the ring of algebraic integers (see [10, Theorem 3.9]). We prove that the two positive integers, i.e., the Casimir number of $\mathcal{C}$ and the Frobenius-Schur exponent of $\mathcal{C}$, have the same prime factors. This may be considered as an integer version of the Cauchy theorem for a spherical fusion category.

Linchenko and Montgomery [50] generalized the classical Frobenius-Schur indicators from group-theoretic result to the setting of a semisimple involutory Hopf algebra H. They also defined higher Frobenius-Schur indicators $\nu_n(V)$ by using idempotent integral Λ of H, namely, $\nu_n(V) = \chi_V(\Lambda_{(1)} \cdots \Lambda_{(n)})$, where $n \geqslant 1$ and χ_V is the character afforded by finite dimensional H-module V. The higher Frobenius-Schur indicators were later extensively studied by Kashina, Sommerhäuser and Zhu

[43] for semisimple Hopf algebras over an algebraically closed field of characteristic zero, and by Ng and Schauenburg [63] for semisimple quasi-Hopf algebras over the field $\mathbb{C}$ of complex numbers. The notations of higher Frobenius-Schur indicators have been generalized to objects of a pivotal category (see [62, 64]).

The notations of higher Frobenius-Schur indicators for semisimple Hopf algebras over a field of positive characteristic seem not to be considered (except for those semisimple involutory Hopf algebras). For a semisimple Hopf algebra H with antipode S over a field $\Bbbk$ of positive characteristic p, it is known that $S^2(h) = \mathbf{u}h\mathbf{u}^{-1}$, where $h \in H$ and $\mathbf{u}$ is a unit of H (see [70, Theorem 5(a)]). This induces a functorial isomorphism $j_{\mathbf{u}} : id \to (-)^{**}$ of the representation category H-mod. If we define the n-th Frobenius-Schur indicator along the lines of [62] to be the trace of a certain $\Bbbk$-linear operator associated with the functorial isomorphism j_u, then the Frobenius-Schur indicator defined here depends on the choice of $\mathbf{u}$. Hence, it is not an invariant of the monoidal category H-mod. Even if a priori the Frobenius-Schur indicator depends on $\mathbf{u}$, with a good choice of $\mathbf{u}$ it may be an invariant of the monoidal category H-mod.

For a finite dimensional semisimple Hopf algebra H over an algebraically closed field $\Bbbk$ of characteristic p, under the condition that $\mathbf{u} := S(\Lambda_{(2)})\Lambda_{(1)}$ is invertible, we obtain the formula $S^2(h) = \mathbf{u}h\mathbf{u}^{-1}$ for $h \in H$. This gives a functorial isomorphism $j_{\mathbf{u}} : id \to (-)^{**}$ of the representation category H-mod. As the element $\mathbf{u}$ is not necessarily a group-like element, the functorial isomorphism $j_{\mathbf{u}}$ is not necessarily a tensor isomorphism. In other words, the monoidal category H-mod is not necessarily pivotal with respect to the structure $j_{\mathbf{u}}$. Even though, using the functorial isomorphism $j_{\mathbf{u}}$ we may still define the n-th Frobenius-Schur indicator $\nu_n(V)$ of an H-module V to be the trace of a certain $\Bbbk$-linear operator as Ng and Schauenburg did in [62].

It may be the best choice of the functorial isomorphism $j_{\mathbf{u}}$ for the definition of higher Frobenius-Schur indicators in positive characteristic. Indeed, by comparison with the case of characteristic 0, we shall see that many properties of a Frobenius-Schur indicator holding in characteristic 0 are preserved in positive characteristic. For example, similarly to the case of characteristic 0, the n-th Frobenius-Schur indicator $\nu_n(V)$ defined here can be entirely described in terms of the integral Λ of H and the character χ_V of V, namely, we have $\nu_n(V) = \chi_V(\mathbf{u}^{-1}\Lambda_{(1)} \cdots \Lambda_{(n)})$. This formula does not depend on the choice of the nonzero integral Λ and it recovers the original formula when the characteristic of $\Bbbk$ is 0. Similarly to the case of characteristic 0, by replacing V with the regular representation H, we reconstruct the n-th indicator of H, a notation defined by the trace of the map $S \circ P_{n-1}$ (see [42, 43]). Similarly to the case of characteristic 0, V and its dual V^* have the same higher Frobenius-Schur indicators. Similarly to the case of characteristic 0, the n-th Frobenius-Schur indicator $\nu_n(V)$ defined here is a monoidal invariant of the representation category H-mod for any $n \in \mathbb{Z}$ and any finite dimensional H-module V.

As mentioned before, for a semisimple Hopf algebra H with antipode S over a field $\Bbbk$, it is known that S^2 is an inner automorphism of H. Here an inner automorphism is understood to be the conjugation by an invertible element of H. If the ground field $\Bbbk$ has positive characteristic p, whether or not S^2 can be given by conjugation with a group-like element is not completely solved (this problem is closely related to the Kaplansky's fifth conjecture). However, such a Hopf algebra H can be embedded into another finite dimensional Hopf algebra $H\#\Bbbk G$, namely, the smash product of H and a group algebra $\Bbbk G$, in which the square of the antipode is the conjugation with a group-like element. We refer to [48, 55, 76] for such Hopf algebras and related researches.

If H is a semisimple involutory Hopf algebra, namely, a semisimple Hopf algebra with $S^2 = id$, the smash product Hopf algebra $H\#\Bbbk G$ considered here is nothing but the usual tensor product Hopf algebra $H \otimes \Bbbk G$. In this case, the representations of $H \otimes \Bbbk G$ can be stemmed directly from the representations of H and those of $\Bbbk G$. Also, the Grothendieck algebra of $H \otimes \Bbbk G$ is the usual tensor product of the Grothendieck algebra of H and that of $\Bbbk G$. However, if H is not necessarily involutory (although the Kaplansky's fifth conjecture states that a semisimple Hopf algebra is necessarily involutory), it is interesting to establish the relationship between the Grothendieck algebra of $H\#\Bbbk G$ and that of H.

Let H be a semisimple Hopf algebra over an algebraically closed field $\Bbbk$ of positive characteristic p. With the conditions that $p > \dim_{\Bbbk}(H)^{1/2}$, $p \nmid 2\dim_{\Bbbk}(H)$ and G is a cyclic group of order $2\dim_{\Bbbk}(H)$, we are able to investigate the Grothendieck algebra of $H\#\Bbbk G$. More explicitly, we endow with a new multiplication operator $\star$ on the Grothendieck algebra $G_{\Bbbk}(H)$ of H so as to obtain a new algebra $(G_{\Bbbk}(H), \star)$. This algebra $(G_{\Bbbk}(H), \star)$ is nothing but the usual Grothendieck algebra $(G_{\Bbbk}(H), *)$ if H is involutory. We show that the Grothendieck algebra $(G_{\Bbbk}(H\#\Bbbk G), *)$ of $H\#\Bbbk G$ is isomorphic to $(G_{\Bbbk}(H), *)^{\oplus \frac{n}{2}} \bigoplus (G_{\Bbbk}(H), \star)^{\oplus \frac{n}{2}}$ as algebras, where $n = 2\dim_{\Bbbk}(H)$. Moreover, we find a fusion subcategory $\mathcal{C}$ of $H\#\Bbbk G$-mod with the Grothendieck algebra $(G_{\Bbbk}(\mathcal{C}), *)$ being isomorphic to $(G_{\Bbbk}(H), *) \bigoplus (G_{\Bbbk}(H), \star)$. In view of this, the Grothendieck algebra $(G_{\Bbbk}(H\#\Bbbk G), *)$ is isomorphic to the direct sum $(G_{\Bbbk}(\mathcal{C}), *)^{\oplus \frac{n}{2}}$.

Note that higher Frobenius-Schur indicators are those invariants induced from the Sweedler power maps on integrals. There should exist some other invariants which can be deduced from the Sweedler power maps on integrals. To investigate these invariants, we denote by P_n and P_n^J the n-th Sweedler power maps of the Hopf algebra H and the twisted Hopf algebra H^J respectively, and describe the relationship between the images of the two maps on the left integral Λ of H. We find the equality $P_n^J(\Lambda) = TP_n(\Lambda)$ for $n \in \mathbb{Z}$, where T is an invertible element of H associated to the twist J. There are some monoidal invariants that can be derived from this equality. For example, we show that the homogeneous polynomials vanishing on some $P_n(\Lambda)$ are invariants of the representation category H-mod. As an application, we show that the representation categories of 12-dimensional pointed non-semisimple Hopf algebras classified in [59] are mutually inequivalent as monoidal categories.

If H is unimodular, then $T = 1$ and hence $P_n^J(\Lambda) = P_n(\Lambda)$ for $n \in \mathbb{Z}$, namely, $P_n(\Lambda)$ is invariant under twisting. If, moreover, H is semisimple and Λ is idempotent, then all polynomials vanishing on some $P_n(\Lambda)$ are monoidal invariants of the representation category H-mod. We use this approach to distinguish the representation categories K_8-mod, $\Bbbk Q_8$-mod and $\Bbbk D_4$-mod, whereas these categories are fusion categories obeying the same fusion rules, we refer to [63, 81] for other approaches to distinguish these fusion categories.

Note that these $P_n(\Lambda)$ are dependent on the choice of a left integral Λ of H. Thus, the values that characters of finite dimensional H-modules take on $P_n(\Lambda)$ are not invariants. In view of this, we turn to consider the ratio $\chi_V(P_n(\Lambda))/\chi_W(P_m(\Lambda))$ for finite dimensional H-modules V and W. When H is unimodular and $\chi_W(P_m(\Lambda)) \neq 0$, we show that this ratio is a monoidal invariant of H-mod. This invariant can be regarded as a generalization of a higher Frobenis-Schur indicator defined for any semisimple Hopf algebra. Indeed, if H is semisimple and W is chosen to be the trivial H-module $\Bbbk$, then the above ratio becomes $\chi_V(P_n(\Lambda/\varepsilon(\Lambda)))$, which is the n-th Frobenius-Schur indicator of V (see [43, 50]).

We further describe a relationship between the right integrals λ of H^* and λ^J of $(H^J)^*$. We show that $\lambda^J = \lambda \leftharpoonup S^2(R^{-1})S(Q_J^{-1})Q_J$, where R and Q are invertible elements of H associated to the twist J. This generalizes the result of [1, Theorem 3.4] for a unimodular Hopf algebra. We use this formula to give a uniform proof of the remarkable result which says that the n-th indicator $\nu_n(H)$ of H is a monoidal invariant of H-mod for any $n \in \mathbb{Z}$. We also use this formula to give an alternative proof of the known result that the Killing form of the Hopf algebra H is invariant under twisting. This gives rise to a result that the dimension of the Killing radical of H is a monoidal invariant of H-mod.

Let us give an outline of the book. Chapters 1—3 deal with the Green rings and the stable Green rings of Hopf algebras, Chapters 4—5 deal with the Casimir numbers of Hopf algebras and fusion categories, Chapter 6 deals with higher Frobenius-Schur indicators of representations of semisimple Hopf algebras in positive characteristic, Chapter 7 deals with the Grothendieck algebras of certain smash product semisimple Hopf algebras, while Chapter 8 deals with certain polynomial invariants on representation categories of finite dimensional Hopf algebras. Next, we will describe the content of the various chapters in more details.

In Chapter 1, we study the Green ring $r(H)$ of a finite dimensional Hopf algebra H by virtue of bilinear forms introduced, e.g., in [8, 65, 94]. We follow the approach of [8] and impose some bilinear forms on the Green ring $r(H)$. We show that two of these bilinear forms are both non-degenerate and they are essentially the same up to a unit of $r(H)$. If H is of finite representation type, the Green ring $r(H)$ can be endowed with an associative and non-degenerate bilinear form $(-,-)$ determined by dimensions of morphism spaces, and hence $r(H)$ is a Frobenius algebra over $\mathbb{Z}$ with a pair of dual bases associated to almost split sequences of H-modules. We use the non-degenerate bilinear form $(-,-)$ to give several one-sided ideals of $r(H)$ and these ideals provide a little more information about the Jacobson radical and central primitive idempotents of $r(H)$. It is known that the Grothendieck ring $G_0(H)$ of H is a quotient ring of $r(H)$. We describe this quotient ring clearly: $r(H)/\mathcal{P}^{\perp} \cong G_0(H)$, where $\mathcal{P}^{\perp}$ is orthogonal to the projective ideal $\mathcal{P}$ with respect to the form $(-,-)$.

In Chapter 2, we pay our attention to the Green ring $r(H)$ of a finite dimensional spherical Hopf algebra H. The category H-mod of finite dimensional left H-modules is a spherical category [2]. This enables us to consider the quotient of $r(H)$ modulo all H-modules of quantum dimension zero. This quotient ring is called the Benson-Carlson quotient ring by Archer [3] as it was first introduced and studied by Benson and Carlson [7]. The Benson-Carlson quotient ring of $r(H)$ considered here has some special properties: it can be realized as the Green ring of a factor category of H-mod by all negligible morphisms. In particular, if H is of finite representation type, this Benson-Carlson quotient ring can be endowed with group-like algebra and bi-Frobenius algebra structure.

In Chapter 3, we use a bilinear form to study the stable Green ring of a finite dimensional Hopf algebra H. The Green ring of the stable category H-<u>mod</u> of H is called the stable Green ring of H, denoted by $r_{st}(H)$. As the stable category H-

mod is a quotient category of H-mod, the stable Green ring $r_{st}(H)$ is shown to be a quotient ring of the Green ring $r(H)$, namely, $r_{st}(H) \cong r(H)/\mathcal{P}$. This isomorphism enables us to define a new bilinear form $[-,-]_{st}$ on $r_{st}(H)$ which is induced from the form $(-,-)$ on $r(H)$. The form $[-,-]_{st}$ is associative but degenerate in general. We determine the left and right radicals of the form $[-,-]_{st}$ respectively, and give several equivalent conditions for the non-degeneracy of the form. Under the assumption that H is of finite representation type, the Green ring $r(H)$ is commutative and the form $[-,-]_{st}$ is non-degenerate, we show that the Jacobson radical of $r(H)$ is equal to $\mathcal{P} \cap \mathcal{P}^{\perp}$ if and only if the Grothendieck ring $G_0(H)$ is semiprime. If H is of finite representation type and the form $[-,-]_{st}$ is non-degenerate, the complexified stable Green algebra is a Frobenius algebra, we show that it is also a group-like algebra, and hence a bi-Frobenius algebra.

In Chapter 4, if the Hopf algebra H is of finite representation type, the Green ring $r(H)$ is associated with a non-negative integer, called the Casimir number of $r(H)$. This number, denoted by m_H, is a monoidal invariant of H-mod. The Casimir number m_H can be used to determine whether or not the Green ring $r(H)$, or the Green algebra $r(H) \otimes_{\mathbb{Z}} K$ over a field K is semisimple. It turns out that $r(H) \otimes_{\mathbb{Z}} K$ is semisimple if and only if the Casimir number m_H is not zero in K. For the Green ring $r(H)$ itself, $r(H)$ is semisimple if and only if the Casimir number m_H is not zero. If the Green ring $r(H)$ is commutative, then the Jacobson radical of $r(H)$ is the set of all nilpotent elements of $r(H)$. Thus, we give a characterization of a commutative Green ring without nonzero nilpotent elements. As an application, we compute the Casimir number $m_{\Bbbk G}$ of the Green ring of $\Bbbk G$, where G is a cyclic group of order p and $\Bbbk$ is an algebraically closed field of characteristic p. By a straightforward computation, we find that $m_{\Bbbk G} = 2p^2$. This shows that the Green ring $r(\Bbbk G)$ is semisimple, which is exactly a result of [34]. Moreover, $r(\Bbbk G) \otimes_{\mathbb{Z}} K$ is semisimple if and only if the characteristic of K is not equal to 2 or p. In the case where K is of characteristic 2 or p, we use the factorizations of Dickson polynomials to describe the Jacobson radical of $r(\Bbbk G) \otimes_{\mathbb{Z}} K$ explicitly.

In Chapter 5, we turn to consider the Casimir numbers of semisimple Hopf algebras and fusion categories. Let H be a finite dimensional semisimple and cosemisimple Hopf algebra over $\Bbbk$. If the Grothendieck ring $G_0(H)$ of H is commutative, we show that the Casimir number m_H and the dimension $\dim_{\Bbbk}(H)$ have the same prime factors. Applying to the Drinfeld double $D(H)$ of H, we obtain that the Casimir number $m_{D(H)}$ and the dimension $\dim_{\Bbbk}(H)$ share the same prime factors. For the Casimir number of a fusion category, we reveal a relationship between the Casimir number $m_{\mathcal{C}}$ of a fusion category $\mathcal{C}$ and the Casimir number $m_{\widetilde{\mathcal{C}}}$ of the pivotalization $\widetilde{\mathcal{C}}$. We show that the former is a factor of the latter. This gives a result that any non-degenerate fusion category over a field $\Bbbk$ has a nonzero Casimir number in $\Bbbk$. For the case that $\mathcal{C}$ is spherical over the field $\mathbb{C}$ of complex numbers, the Casimir number of $\mathcal{C}$ and the Frobenius-Schur exponent of $\mathcal{C}$ share the same prime factors. This may be considered as an integer version of the Cauchy theorem for a spherical fusion category.

In Chapter 6, we consider higher Frobenius-Schur indicators for representations of a finite dimensional semisimple Hopf algebra H over an algebraically closed field $\Bbbk$ of characteristic p. Under the condition that $\mathbf{u} = S(\Lambda_{(2)})\Lambda_{(1)}$ is invertible, we investigate some properties of the element $\mathbf{u}$ and obtain a formula for S^2, i.e., $S^2(h) = \mathbf{u}h\mathbf{u}^{-1}$ for $h \in H$. Using this formula, we generalize the notations of higher Frobenius-Schur indicators from characteristic 0 case to characteristic p case and find that the indicators defined here share some common properties with the ones defined over a field of characteristic 0. Especially, these higher Frobenius-Schur indicators are monoidal invariants of the representation category H-mod.

In Chapter 7, when H is a semisimple Hopf algebra over an algebraically closed field $\Bbbk$ of positive characteristic p with $p > \dim_{\Bbbk}(H)^{1/2}$, $p \nmid 2\dim_{\Bbbk}(H)$ and G is a cyclic group of order $2\dim_{\Bbbk}(H)$, we describe all non-isomorphic irreducible representations of the smash product semisimple Hopf algebra $H\#\Bbbk G$ and establish a relationship between the Grothendieck algebra of $H\#\Bbbk G$ and that of H. It turns out that the Grothendieck algebra $(G_{\Bbbk}(H\#\Bbbk G), *)$ of $H\#\Bbbk G$ is isomorphic as algebras to $(G_{\Bbbk}(H), *)^{\oplus \frac{n}{2}} \bigoplus (G_{\Bbbk}(H), \star)^{\oplus \frac{n}{2}}$, where $n = 2\dim_{\Bbbk}(H)$ and $\star$ is a new multiplication operator on the Grothendieck algebra $(G_{\Bbbk}(H), *)$ of H.

In Chapter 8, we concern those monoidal invariants raised from the images of the Sweedler power maps on integrals. We first investigate a relationship between the images of the Sweedler power maps P_n and P_n^J on a left integral Λ of a finite dimensional Hopf algebra H. This relationship reveals certain polynomial invariants of the representation category H-mod. Using these polynomial invariants we can distinguish the representation categories of some Hopf algebras, including all 12-dimensional pointed non-semisimple Hopf algebras, the 8-dimensional semisimple Hopf algebras K_8, $\Bbbk Q_8$ and $\Bbbk D_4$. We also show that a right integral λ^J of $(H^J)^*$ can be expressed in terms of a right integral λ of H^*. This enables us to give a uniform proof of the remarkable result which says that the n-th indicator $\nu_n(H)$ for any $n \in \mathbb{Z}$ is a monoidal invariant of H-mod. This also enables us to give an alternative proof of the known result that the Killing form of the Hopf algebra H is invariant under twisting.

This book is based on the author's post-doctoral research report and the papers [88–91]. I am indebted to my co-supervisor Professor Gongxiang Liu for his guidance. I am grateful to Professor Yinhuo Zhang for his continued guidance, discussions and support during my stay in Hasselt University. I also want to thank Professors Libin Li and Huixiang Chen for their advice and encouragement.

Zhihua Wang
August, 2023

Contents

Chapter 1 The Green Rings of Hopf Algebras

Throughout this chapter, H is an arbitrary finite dimensional Hopf algebra over an algebraically closed field $\Bbbk$. The category of finite dimensional left H-modules is denoted by H-mod. All H-modules considered here are objects in H-mod. The tensor product $\otimes$ stands for $\otimes_\Bbbk$. In this chapter, we characterize when the trivial H-module $\Bbbk$ is a direct summand of the decomposition of a tensor product of any two indecomposable H-modules. We then endow with some bilinear forms on the Green ring $r(H)$ and investigate the relationships among these bilinear forms. If the Hopf algebra H is of finite representation type, we show that the Green ring $r(H)$ is a Frobenius algebra over the ring $\mathbb{Z}$ of integers with a pair of dual bases associated to almost split sequences of H-modules. We will introduce the notation $N \mid M$ to mean that N is (isomorphic to) a summand of M.

1.1 Hopf algebras

Recall that an *algebra* is a vector space A over a field $\Bbbk$ with a linear multiplication map $m : A \otimes A \to A$, and a unit given by a linear unit map $u : \Bbbk \to A$, such that the following diagrams commute:

$$
\begin{array}{ccc}
A \otimes A \otimes A & \xrightarrow{\;m \otimes id\;} & A \otimes A \\
{\scriptstyle id \otimes m}\big\downarrow & & \big\downarrow{\scriptstyle m} \\
A \otimes A & \xrightarrow{\;m\;} & A
\end{array}
\qquad\qquad
\begin{array}{ccc}
& A \otimes A & \\
{\scriptstyle u \otimes id}\nearrow & \big\downarrow{\scriptstyle m} & \nwarrow{\scriptstyle id \otimes u} \\
\Bbbk \otimes A & & A \otimes \Bbbk \\
{\scriptstyle \simeq}\searrow & \big\downarrow & \swarrow{\scriptstyle \simeq} \\
& A &
\end{array}
$$

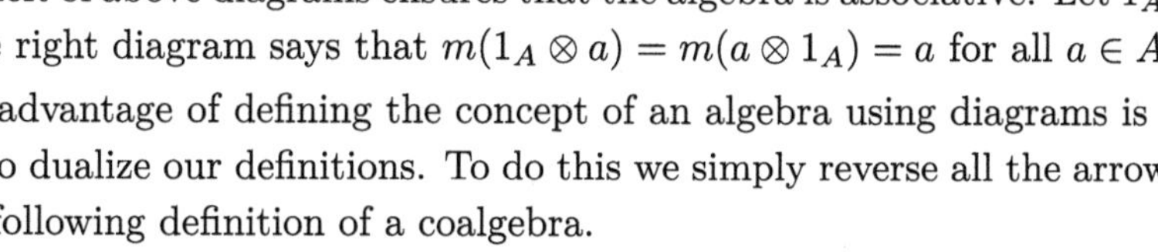

The left of above diagrams ensures that the algebra is associative. Let $1_A := u(1)$, then the right diagram says that $m(1_A \otimes a) = m(a \otimes 1_A) = a$ for all $a \in A$.

The advantage of defining the concept of an algebra using diagrams is that it is simply to dualize our definitions. To do this we simply reverse all the arrows. So we get the following definition of a coalgebra.

A *coalgebra* is a vector space C over a field $\Bbbk$ together with two linear maps, one is *comultiplication* $\Delta : C \to C \otimes C$ and another is *counit* $\varepsilon : C \to \Bbbk$ such that the following diagrams commute:

$$
\begin{array}{ccc}
C & \xrightarrow{\;\Delta\;} & C \otimes C \\
\Delta \downarrow & & \downarrow \Delta \otimes id \\
C \otimes C & \xrightarrow[id \otimes \Delta]{} & C \otimes C \otimes C
\end{array}
\qquad\qquad
\begin{array}{c}
C \\
\Bbbk \otimes C \xleftarrow{\;\simeq\;} \;\Delta\; \xrightarrow{\;\simeq\;} C \otimes \Bbbk \\
\xleftarrow{\varepsilon \otimes id} \; C \otimes C \; \xrightarrow{id \otimes \varepsilon}
\end{array}
$$

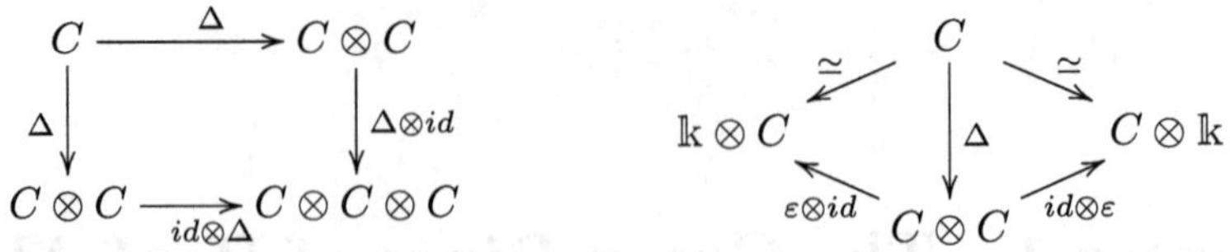

We usually denote $\Delta(a) = a_{(1)} \otimes a_{(2)}$ for $a \in C$, where the sum sign is omitted. The left of the above diagrams expresses the so called coassociative property of the comultiplication Δ, i.e., $(id \otimes \Delta) \circ \Delta = (\Delta \otimes id) \circ \Delta$, or equivalently,

$$
a_{(1)(1)} \otimes a_{(1)(2)} \otimes a_{(2)} = a_{(1)} \otimes a_{(2)(1)} \otimes a_{(2)(2)}.
$$

By convention, both sides of the above equation are identified with $a_{(1)} \otimes a_{(2)} \otimes a_{(3)}$.

For a coalgebra C, an element $g \in C$ which satisfies $\Delta(g) = g \otimes g$ and $\varepsilon(g) = 1$ is called a *group-like element* of C. For two coalgebras C and D, a $\Bbbk$-linear map $\alpha : C \to D$ is called a *coalgebra morphism* if the following diagrams commute:

$$
\begin{array}{ccc}
C & \xrightarrow{\;\Delta_C\;} & C \otimes C \\
\alpha \downarrow & & \downarrow \alpha \otimes \alpha \\
D & \xrightarrow[\Delta_D]{} & D \otimes D
\end{array}
\qquad\qquad
\begin{array}{ccc}
C & \xrightarrow{\;\varepsilon_C\;} & \\
\alpha \downarrow & & \Bbbk \\
D & \xrightarrow[\varepsilon_D]{} &
\end{array}
$$

A *bialgebra* is a vector space that is both an algebra and a coalgebra. Further, the algebra and coalgebra structures must be compatible. The compatibility is ensured by requiring either of the following equivalent conditions:

(1) Δ and ε must be algebra morphisms.

(2) m and u must be coalgebra morphisms.

Given an algebra A, a coalgebra C and two $\Bbbk$-linear maps $\alpha, \beta : C \to A$, we can get a new map from C to A as follows,

$$
C \xrightarrow{\Delta} C \otimes C \xrightarrow{\alpha \otimes \beta} A \otimes A \xrightarrow{m} A.
$$

The linear map obtained in this way is called the *convolution* of α and β, denoted by $\alpha * \beta$, namely,

$$
\alpha * \beta = m \circ (\alpha \otimes \beta) \circ \Delta.
$$

Now an *antipode* is a $\Bbbk$-linear map S from a bialgebra H to itself, which satisfies

$$
id * S = S * id = u \circ \varepsilon.
$$

The antipode S is indeed the convolution inverse to the identity id of H. If it exists, it is necesarilly unique. A *Hopf algebra* is a bialgebra equipped with an antipode. For more details about Hopf algebras, we refer to [44, 56, 80].

Let H denote a finite dimensional Hopf algebra over a field $\Bbbk$, with comultiplication Δ, counit ε and antipode S. A *left integral* of H is an element $\Lambda \in H$ such that $h\Lambda = \varepsilon(h)\Lambda$ for all $h \in H$; similarly $\Gamma \in H$ is a *right integral* of H if $\Gamma h = \varepsilon(h)\Gamma$ for all $h \in H$. It is known that the space of left (resp. right) integrals is one dimensional. If the left integral space coincides with the right one, then H is called *unimodular*. In particular, any finite dimensional semisimple Hopf algebra is unimodular.

The category H-mod of finite dimensional left H-modules is a monoidal category, where the monoidal structure is given by the comultiplication Δ, namely, if V and W are H-modules, then the tensor product $V \otimes W$ is also an H-module via

$$h \cdot (v \otimes w) := \Delta(h)(v \otimes w) = h_{(1)}v \otimes h_{(2)}w,$$

for $h \in H$, $v \in V$ and $w \in W$.

Recall that the Hom-space $\mathrm{Hom}_{\Bbbk}(V, W)$ is an H-module given by $(hf)(v) = h_{(1)}f(S(h_{(2)})v)$, for $h \in H$, $v \in V$ and $f \in \mathrm{Hom}_{\Bbbk}(V, W)$. In the special case where W is the trivial module $\Bbbk$, then $V^* := \mathrm{Hom}_{\Bbbk}(V, \Bbbk)$ is an H-module given by $(hf)(v) = f(S(h)v)$, for $h \in H$, $v \in V$ and $f \in V^*$. In general, V^{**} is not isomorphic to V as H-modules unless S^2 is an inner automorphism of H.

If V is a finite dimensional H-module, then V is also called a *representation* of H via the algebra homomorphism $\rho_V : H \to \mathrm{End}_{\Bbbk}(V)$ given by $\rho_V(h)(v) = hv$ for $h \in H$ and $v \in V$. We will make no distinction between the two notations. The *character* of V is the map $\chi_V : H \to \Bbbk$ given by $\chi_V(h) = \mathrm{tr}(\rho_V(h))$ for $h \in H$. In particular, the dual module V^* has the character $\chi_{V^*} = \chi_V \circ S$.

The *evaluation* of V is the morphism $\mathrm{ev}_V : V^* \otimes V \to \Bbbk$ given by $\mathrm{ev}_V(f \otimes v) = f(v)$. The *coevaluation* of V is the morphism $\mathrm{coev}_V : \Bbbk \to V \otimes V^*$ defined by $\mathrm{coev}_V(1) = \sum_i v_i \otimes v_i^*$, where $\{v_i\}$ and $\{v_i^*\}$ are bases of V and V^* respectively such that $v_i^*(v_j) = \delta_{i,j}$, where $\delta_{i,j}$ is the Kronecker symbol.

The following notations can be found in [4]. Let M and N be two H-modules. The morphism $f : M \to N$ is called a *split monomorphism* if there exists $g : N \to M$ such that $g \circ f = id_M$, and $f : M \to N$ is called a *split epimorphism* if there exists $g : N \to M$ such that $f \circ g = id_N$. The morphism $f : M \to N$ is called *left almost split* if f is not a split monomorphism and for any H-module X, if there is $g : M \to X$ with g not a split monomorphism, then there is $h : N \to X$ such that $h \circ f = g$. Dually, $f : M \to N$ is called *right almost split* if f is not a split epimorphism and for any H-module Y, if there is $g : Y \to N$ with g not a split epimorphism, then there is $h : Y \to M$ such that $f \circ h = g$. The morphism $f : M \to N$ is called *left minimal* if for all $h : N \to N$ with $h \circ f = f$, then h is an isomorphism. Dually, $f : M \to N$ is called *right minimal* if for all $h : M \to M$ with $f \circ h = f$, then h is an isomorphism. A short exact sequence $0 \to X \xrightarrow{f} M \xrightarrow{g} Y \to 0$ is called *almost split* if f is both left minimal and left almost split and g is both right minimal and right almost split.

1.2 Quantum traces of morphisms

In the study of the Green ring $r(H)$ of a Hopf algebra H, one of difficult problems is to determine whether or not the trivial module $\Bbbk$ appears in the decomposition of the tensor product $X \otimes Y$ for indecomposable modules X and Y. This problem has already been solved in the case of group algebras by Benson and Carlson [7, Theorem 2.1], in the case of involutory Hopf algebras in terms of splitting trace modules [35], and in the case of Hopf algebras with the square of antipode being inner [95, Theorem 2.4]. Motivated by these works, in this section we use quantum traces of morphisms of H-modules to characterize when the trivial module $\Bbbk$ is a direct summand of the decomposition of a tensor product of indecomposable modules. Consequently, we answer the question raised by Cibils [18, Remark 5.8]. In particular, we apply techniques from [35, 95] to determine whether or not the trivial module $\Bbbk$ appears in the decomposition of the tensor product $X \otimes X^*$ (resp. $X^* \otimes X$) for any indecomposable module X. Most results stated in this section are useful for next sections.

The *left quantum trace* of $\theta \in \mathrm{Hom}_H(X, X^{**})$ is defined to be the following composition:

$$\mathrm{Tr}^L_X(\theta): \Bbbk \xrightarrow{\mathrm{coev}_X} X \otimes X^* \xrightarrow{\theta \otimes id_{X^*}} X^{**} \otimes X^* \xrightarrow{\mathrm{ev}_{X^*}} \Bbbk. \tag{1.1}$$

Similarly, the *right quantum trace* of $\theta \in \mathrm{Hom}_H(X^{**}, X)$ is defined to be

$$\mathrm{Tr}^R_X(\theta): \Bbbk \xrightarrow{\mathrm{coev}_{X^*}} X^* \otimes X^{**} \xrightarrow{id_{X^*} \otimes \theta} X^* \otimes X \xrightarrow{\mathrm{ev}_X} \Bbbk. \tag{1.2}$$

Since $\mathrm{End}_H(\Bbbk) \cong \Bbbk$, both $\mathrm{Tr}^L_X(\theta)$ and $\mathrm{Tr}^R_X(\theta)$ are elements in $\Bbbk$.

Remark 1.2.1 *Applying the duality functor* $*$ *to* (1.1) *and* (1.2) *respectively, one obtains that* $\mathrm{Tr}^L_X(\theta) = \mathrm{Tr}^R_{X^*}(\theta^*)$ *and* $\mathrm{Tr}^R_X(\theta) = \mathrm{Tr}^L_{X^*}(\theta^*)$.

Remark 1.2.2 *Let P be a projective H-module.*

(1) *If H is not semisimple, then* $\mathrm{Tr}^L_P(\theta) = 0$ *for any* $\theta \in \mathrm{Hom}_H(P, P^{**})$. *Otherwise, the morphism* coev_P *is a split monomorphism by* (1.1). *In this case,* $\Bbbk | P \otimes P^*$. *It follows that $\Bbbk$ is projective, and hence H is semisimple, a contradiction. Similarly, if H is not semisimple, then* $\mathrm{Tr}^R_P(\theta) = 0$ *for any* $\theta \in \mathrm{Hom}_H(P^{**}, P)$.

(2) *If H is involutory, i.e.,* $S^2 = id$, *then the map* $\theta : P \to P^{**}$ *given by* $\theta(x)(f) = f(x)$ *for* $x \in P$ *and* $f \in P^*$ *is an H-module isomorphism. In this case,*

$$\mathrm{Tr}^L_P(\theta) = \mathrm{Tr}^R_P(\theta^{-1}) = \dim_{\Bbbk}(P).$$

This implies that an involutory Hopf algebra over a field $\Bbbk$ *of characteristic* 0 *is semisimple* (*the converse is also true, see* [46]). *In the case the characteristic of* $\Bbbk$ *is* $p > 0$ *and* H *is not semisimple, then* $p \mid \dim_{\Bbbk}(P)$, *giving a result of Lorenz* [51, *Theorem* 2.3 (*b*)].

Lemma 1.2.3 *Let* X *be an indecomposable* H*-module.*

(1) *For any* $\theta \in \mathrm{Hom}_H(X, X^{**})$, *if* $\mathrm{Tr}_X^L(\theta) \neq 0$, *then* θ *is an isomorphism.*

(2) *For any* $\theta \in \mathrm{Hom}_H(X^{**}, X)$, *if* $\mathrm{Tr}_X^R(\theta) \neq 0$, *then* θ *is an isomorphism.*

Proof We only prove Part (1) and the proof of Part (2) is similar. For any integer $m > 0$, the m-th power of the duality functor $*$ on X is denoted by X^{*m}. If $\{x_i\}$ is a basis of X, we denote by $\{x_i^{*m}\}$ the basis of X^{*m} dual to the basis $\{x_i^{*m-1}\}$ of X^{*m-1}, i.e., $\langle x_i^{*m}, x_j^{*m-1}\rangle = \delta_{i,j}$. Let $\mathbf{A}$ be the transformation matrix of the morphism $\theta \in \mathrm{Hom}_H(X, X^{**})$ with respect to the bases $\{x_i\}$ and $\{x_i^{**}\}$. It is clear that $\mathrm{Tr}_X^L(\theta) = \mathrm{tr}(\mathbf{A})$, which is the usual trace of the matrix $\mathbf{A}$. Since H is a finite dimensional Hopf algebra, the order of S^2 is finite by Radford's formula on S^4 and the Nichols-Zöller Theorem. Suppose that $S^{2n} = id$. Then the map

$$\mathrm{Id} : X^{*2n} \to X, \quad \sum_i \lambda_i x_i^{*2n} \mapsto \sum_i \lambda_i x_i$$

is an H-module isomorphism and the transformation matrix of the map Id with respect to the basis $\{x_i^{*2n}\}$ of X^{*2n} and the basis $\{x_i\}$ of X is the identity matrix. Consider the following composition:

$$\Theta : X \xrightarrow{\theta} X^{**} \xrightarrow{\theta^{**}} X^{****} \to \cdots \to X^{*2n-2} \xrightarrow{\theta^{*2n-2}} X^{*2n} \xrightarrow{\mathrm{Id}} X.$$

Note that the matrix of the morphism Θ from X to itself with respect to the basis $\{x_i\}$ of X is $\mathbf{A}^n$. Since $\mathrm{End}_H(X)$ is local, the morphism Θ is either nilpotent or isomorphic. If Θ is nilpotent, so is $\mathbf{A}^n$, and hence $\mathbf{A}$ is nilpotent. This implies that $\mathrm{Tr}_X^L(\theta) = \mathrm{tr}(\mathbf{A}) = 0$, a contradiction. Thus, Θ is an isomorphism, and so is the morphism θ. □

The following lemma follows from [5, Lemma 2.1.6]. We present it here because it will be used later.

Lemma 1.2.4 *For* H*-modules* X, Y *and* Z, *we have the following canonical isomorphisms functorial in* X, Y *and* Z:

(1) $\Phi_{X,Y,Z} : \mathrm{Hom}_H(X \otimes Y, Z) \to \mathrm{Hom}_H(X, Z \otimes Y^*)$, $\Phi_{X,Y,Z}(\alpha) = (\alpha \otimes id_{Y^*}) \circ (id_X \otimes \mathrm{coev}_Y)$.

(2) $\Psi_{X,Y,Z} : \mathrm{Hom}_H(X, Y \otimes Z) \to \mathrm{Hom}_H(Y^* \otimes X, Z)$, $\Psi_{X,Y,Z}(\gamma) = (\mathrm{ev}_Y \otimes id_Z) \circ (id_{Y^*} \otimes \gamma)$.

The inverse maps of $\Phi_{X,Y,Z}$ and $\Psi_{X,Y,Z}$, respectively, are

$$\Phi^{-1}_{X,Y,Z}(\beta) = (id_Z \otimes \mathrm{ev}_Y) \circ (\beta \otimes id_Y),$$

for $\beta \in \mathrm{Hom}_H(X, Z\otimes Y^*)$; and

$$\Psi^{-1}_{X,Y,Z}(\delta) = (id_Y \otimes \delta)\circ(\mathrm{coev}_Y \otimes id_X),$$

for $\delta \in \mathrm{Hom}_H(Y^*\otimes X, Z)$. The two canonical isomorphisms satisfy the following properties:

Proposition 1.2.5 *Let X be an indecomposable H-module. For any H-module Y, we have*

(1) *The canonical isomorphism* $\mathrm{Hom}_H(Y\otimes X^*, \Bbbk) \xrightarrow{\Phi_{Y,X^*,\Bbbk}} \mathrm{Hom}_H(Y, X^{**})$ *preserves split epimorphisms.*

(2) *The canonical isomorphism* $\mathrm{Hom}_H(Y, X) \xrightarrow{\Psi_{Y,X,\Bbbk}} \mathrm{Hom}_H(X^*\otimes Y, \Bbbk)$ *reflects split epimorphisms.*

Proof (1) If the morphism $\alpha \in \mathrm{Hom}_H(Y\otimes X^*, \Bbbk)$ is a split epimorphism, there is some $\beta \in \mathrm{Hom}_H(\Bbbk, Y\otimes X^*)$ such that $\alpha\circ\beta = id_\Bbbk$. For the morphism β, there is some $\gamma \in \mathrm{Hom}_H(X, Y)$ such that $\beta = \Phi_{\Bbbk,X,Y}(\gamma)$. Note that $\Phi_{Y,X^*,\Bbbk}(\alpha)\circ\gamma \in \mathrm{Hom}_H(X, X^{**})$. It follows that

$$\begin{aligned}
&\mathrm{Tr}^L_X(\Phi_{Y,X^*,\Bbbk}(\alpha)\circ\gamma)\\
&= \mathrm{ev}_{X^*}\circ((\Phi_{Y,X^*,\Bbbk}(\alpha)\circ\gamma)\otimes id_{X^*})\circ\mathrm{coev}_X\\
&= (id_\Bbbk\otimes \mathrm{ev}_{X^*})\circ(\Phi_{Y,X^*,\Bbbk}(\alpha)\otimes id_{X^*})\circ(\gamma\otimes id_{X^*})\circ(id_\Bbbk\otimes\mathrm{coev}_X)\\
&= \Phi^{-1}_{Y,X^*,\Bbbk}(\Phi_{Y,X^*,\Bbbk}(\alpha))\circ\Phi_{\Bbbk,X,Y}(\gamma)\\
&= \alpha\circ\beta = id_\Bbbk.
\end{aligned}$$

Thus, $\Phi_{Y,X^*,\Bbbk}(\alpha)\circ\gamma$ is an isomorphism by Lemma 1.2.3, and hence the morphism $\Phi_{Y,X^*,\Bbbk}(\alpha)$ is a split epimorphism.

(2) If the morphism $\alpha \in \mathrm{Hom}_H(X^*\otimes Y, \Bbbk)$ is a split epimorphism, there is some $\beta \in \mathrm{Hom}_H(\Bbbk, X^*\otimes Y)$ such that $\alpha\circ\beta = id_\Bbbk$. Note that

$$\Psi^{-1}_{Y,X,\Bbbk}(\alpha)\circ\Psi_{\Bbbk,X^*,Y}(\beta) \in \mathrm{Hom}_H(X^{**}, X).$$

It follows that

$$\begin{aligned}
&\mathrm{Tr}^R_X(\Psi^{-1}_{Y,X,\Bbbk}(\alpha)\circ\Psi_{\Bbbk,X^*,Y}(\beta))\\
&= \mathrm{ev}_X\circ(id_{X^*}\otimes(\Psi^{-1}_{Y,X,\Bbbk}(\alpha)\circ\Psi_{\Bbbk,X^*,Y}(\beta)))\circ\mathrm{coev}_{X^*}\\
&= (\mathrm{ev}_X\otimes id_\Bbbk)\circ(id_{X^*}\otimes\Psi^{-1}_{Y,X,\Bbbk}(\alpha))\circ(id_{X^*}\otimes\Psi_{\Bbbk,X^*,Y}(\beta))\circ(\mathrm{coev}_{X^*}\otimes id_\Bbbk)\\
&= \Psi_{Y,X,\Bbbk}(\Psi^{-1}_{Y,X,\Bbbk}(\alpha))\circ\Psi^{-1}_{\Bbbk,X^*,Y}(\Psi_{\Bbbk,X^*,Y}(\beta))\\
&= \alpha\circ\beta = id_\Bbbk.
\end{aligned}$$

Thus, $\Psi^{-1}_{Y,X,\Bbbk}(\alpha) \circ \Psi_{\Bbbk,X^*,Y}(\beta)$ is an isomorphism by Lemma 1.2.3, and hence the morphism $\Psi^{-1}_{Y,X,\Bbbk}(\alpha)$ is a split epimorphism. $\square$

As an immediate consequence of Proposition 1.2.5, we have the following result:

Corollary 1.2.6 *Let X and Y be two indecomposable H-modules.*

(1) *If $\Bbbk \mid Y \otimes X^*$, then $Y \cong X^{**}$.*

(2) *If $\Bbbk \mid X^* \otimes Y$, then $Y \cong X$.*

Cibils in [18, Remark 5.8] raised the following question: when is the trivial module a direct summand of the tensor product of two indecomposable modules over a finite dimensional Hopf algebra? We are now ready to answer this question using quantum traces.

Theorem 1.2.7 *Let X and Y be two indecomposable H-modules.*

(1) *$\Bbbk \mid Y \otimes X^*$ if and only if there are isomorphisms $f : X \to Y$ and $g : Y \to X^{**}$ such that $\mathrm{Tr}^L_X(g \circ f) \neq 0$.*

(2) *$\Bbbk \mid X^* \otimes Y$ if and only if there are isomorphisms $f : X^{**} \to Y$ and $g : Y \to X$ such that $\mathrm{Tr}^R_X(g \circ f) \neq 0$.*

Proof We only prove Part (1) and the same argument works for Part (2). If $f : X \to Y$ and $g : Y \to X^{**}$ are two isomorphisms such that $\mathrm{Tr}^L_X(g \circ f) \neq 0$, then

$$0 \neq \mathrm{Tr}^L_X(g \circ f) = \mathrm{ev}_{X^*} \circ (g \otimes id_{X^*}) \circ (f \otimes id_{X^*}) \circ \mathrm{coev}_X.$$

This implies that the morphism $(f \otimes id_{X^*}) \circ \mathrm{coev}_X : \Bbbk \to Y \otimes X^*$ is a split monomorphism, and hence $\Bbbk \mid Y \otimes X^*$. Conversely, if $\Bbbk \mid Y \otimes X^*$, there are morphisms $\alpha : \Bbbk \to Y \otimes X^*$ and $\beta : Y \otimes X^* \to \Bbbk$ such that $\beta \circ \alpha = id_{\Bbbk}$. For the morphism α, by Lemma 1.2.4, there is a morphism $f : X \to Y$ such that

$$\alpha = \Phi_{\Bbbk,X,Y}(f) = (f \otimes id_{X^*}) \circ (id_{\Bbbk} \otimes \mathrm{coev}_X).$$

For the morphism β, there is a morphism $g : Y \to X^{**}$ such that

$$\beta = \Phi^{-1}_{Y,X^*,\Bbbk}(g) = (id_{\Bbbk} \otimes \mathrm{ev}_{X^*}) \circ (g \otimes id_{X^*}).$$

Thus, we have

$$\begin{aligned} \mathrm{Tr}^L_X(g \circ f) &= \mathrm{ev}_{X^*} \circ (g \otimes id_{X^*}) \circ (f \otimes id_{X^*}) \circ \mathrm{coev}_X \\ &= (id_{\Bbbk} \otimes \mathrm{ev}_{X^*}) \circ (g \otimes id_{X^*}) \circ (f \otimes id_{X^*}) \circ (id_{\Bbbk} \otimes \mathrm{coev}_X) \\ &= \beta \circ \alpha \\ &= id_{\Bbbk}. \end{aligned}$$

It follows from Lemma 1.2.3 that the composition $g \circ f$ is an isomorphism. Thus, f and g are both isomorphisms. $\square$

Given two H-modules X and Y, one knows little about how to decompose the tensor product $X \otimes Y$ into a direct sum of indecomposable modules. However, it can be seen from the following proposition that there are still some rules that the decomposition should follow.

Proposition 1.2.8 *Let X, Y, M be H-modules with X and M indecomposable.*

(1) If $\Bbbk \mid M \otimes M^*$ and $M \mid X \otimes Y$, then $\Bbbk \mid X \otimes X^*$ and $X \mid M \otimes Y^*$.

(2) If $\Bbbk \mid M^* \otimes M$ and $M \mid Y \otimes X$, then $\Bbbk \mid X^* \otimes X$ and $X \mid Y^* \otimes M$.

Proof (1) We only prove Part (1) and the proof of Part (2) is similar. The conditions $\Bbbk \mid M \otimes M^*$ and $M \mid X \otimes Y$ imply that $\Bbbk \mid X \otimes Y \otimes M^*$. Suppose $Y \otimes M^* \cong \bigoplus_i N_i^*$ for some indecomposable modules N_i. Then there is an indecomposable module N_i such that $\Bbbk \mid X \otimes N_i^*$. By Theorem 1.2.7 (1), we obtain $X \cong N_i \cong N_i^{**}$. It follows that

$$\Bbbk \mid X \otimes N_i^* \cong X \otimes X^*.$$

Note that $\Bbbk \mid M \otimes M^*$ implies that $M \cong M^{**}$. Then $X \cong N_i^{**}$ implies that $X \mid (Y \otimes M^*)^* \cong M \otimes Y^*$, as desired. □

In the rest of this section, H will be a non-semisimple Hopf algebra. We shall take another approach to characterize when the trivial module $\Bbbk$ appears in the decomposition of the tensor product $X^* \otimes X$ (resp. $X \otimes X^*$) for an indecomposable module X. For the special case where the square of the antipode is inner, we refer to [35, 95]. Suppose

$$0 \to \tau(\Bbbk) \to E \xrightarrow{\sigma} \Bbbk \to 0 \tag{1.3}$$

is an almost split sequence ending at the trivial module $\Bbbk$. Tensoring (over $\Bbbk$) the sequence (1.3) with an indecomposable module X, we obtain the following two short exact sequences:

$$0 \to \tau(\Bbbk) \otimes X \to E \otimes X \xrightarrow{\sigma \otimes id_X} X \to 0, \tag{1.4}$$

$$0 \to X \otimes \tau(\Bbbk) \to X \otimes E \xrightarrow{id_X \otimes \sigma} X \to 0. \tag{1.5}$$

We need the following lemma, its proof is straightforward if one applies Lemma 1.2.4.

Lemma 1.2.9 *For H-modules X and Y, the following diagrams are commutative:*

$$\begin{array}{ccc} \mathrm{Hom}_H(Y, X \otimes E) & \xrightarrow{(id_X \otimes \sigma)_*} & \mathrm{Hom}_H(Y, X) \\ {\scriptstyle \Psi_{Y,X,E}} \downarrow & & \downarrow {\scriptstyle \Psi_{Y,X,\Bbbk}} \\ \mathrm{Hom}_H(X^* \otimes Y, E) & \xrightarrow{\sigma_*} & \mathrm{Hom}_H(X^* \otimes Y, \Bbbk) \end{array} \tag{1.6}$$

$$\begin{array}{ccc} \mathrm{Hom}_H(Y\otimes X,E) & \xrightarrow{\sigma_*} & \mathrm{Hom}_H(Y\otimes X,\Bbbk) \\ {\scriptstyle \Phi_{Y,X,E}}\downarrow & & \downarrow{\scriptstyle \Phi_{Y,X,\Bbbk}} \\ \mathrm{Hom}_H(Y,E\otimes X^*) & \xrightarrow{(\sigma\otimes id_{X^*})_*} & \mathrm{Hom}_H(Y,X^*) \end{array} \tag{1.7}$$

Proposition 1.2.10 *Let X be an indecomposable H-module. The following statements are equivalent:*

(1) $\Bbbk \nmid X^* \otimes X$.

(2) *The map* $\mathrm{Hom}_H(X^*\otimes X,E) \xrightarrow{\sigma_*} \mathrm{Hom}_H(X^*\otimes X,\Bbbk)$ *is surjective.*

(3) *The map* $\mathrm{Hom}_H(X,X\otimes E) \xrightarrow{(id_X\otimes\sigma)_*} \mathrm{Hom}_H(X,X)$ *is surjective.*

(4) *The map* $X\otimes E \xrightarrow{id_X\otimes\sigma} X$ *is a split epimorphism.*

(5) *The map* $E\otimes X^* \xrightarrow{\sigma\otimes id_{X^*}} X^*$ *is a split epimorphism.*

Proof (1) $\Leftrightarrow$ (2). If $\Bbbk \nmid X^*\otimes X$, then for any $\alpha \in \mathrm{Hom}_H(X^*\otimes X,\Bbbk)$, the morphism α is not a split epimorphism. Since σ is right almost split from E to $\Bbbk$, there is a morphism β from $X^*\otimes X$ to E such that $\sigma\circ\beta=\alpha$. This implies that σ_* is surjective. Conversely, if the map σ_* is surjective, then $\Bbbk \nmid X^*\otimes X$. Otherwise, by Theorem 1.2.7 (2), there is an isomorphism $\theta : X^{**} \to X$ such that

$$\mathrm{Tr}_X^R(\theta) = id_{\Bbbk}.$$

For the morphism $\mathrm{ev}_X : X^*\otimes X \to \Bbbk$, there is some $\beta \in \mathrm{Hom}_H(X^*\otimes X,E)$ such that $\sigma\circ\beta = \mathrm{ev}_X$ since the map σ_* is surjective. It follows that

$$id_{\Bbbk} = \mathrm{Tr}_X^R(\theta) = \mathrm{ev}_X \circ (id_{X^*}\otimes\theta)\circ \mathrm{coev}_{X^*} = \sigma\circ\beta\circ(id_{X^*}\otimes\theta)\circ\mathrm{coev}_{X^*}.$$

We obtain that the morphism σ is a split epimorphism, a contradiction to the fact that σ is right almost split.

(2) $\Leftrightarrow$ (3). According to the commutative diagram (1.6), we have the following commutative diagram:

$$\begin{array}{ccc} \mathrm{Hom}_H(X,X\otimes E) & \xrightarrow{(id_X\otimes\sigma)_*} & \mathrm{Hom}_H(X,X) \\ {\scriptstyle \Psi_{X,X,E}}\downarrow & & \downarrow{\scriptstyle \Psi_{X,X,\Bbbk}} \\ \mathrm{Hom}_H(X^*\otimes X,E) & \xrightarrow{\sigma_*} & \mathrm{Hom}_H(X^*\otimes X,\Bbbk) \end{array}$$

It follows that σ_* is surjective if and only if $(id_X\otimes\sigma)_*$ is surjective.

(3) $\Leftrightarrow$ (4). If $(id_X\otimes\sigma)_*$ is surjective, for the identity map id_X, there is a morphism $\alpha\in\mathrm{Hom}_H(X,X\otimes E)$ such that $(id_X\otimes\sigma)_*(\alpha)=id_X$, namely, $(id_X\otimes\sigma)\circ\alpha = id_X$. It follows that $id_X\otimes\sigma$ is a split epimorphism. Conversely, if $id_X\otimes\sigma$

is a split epimorphism, there is $\alpha \in \mathrm{Hom}_H(X, X \otimes E)$ such that $(id_X \otimes \sigma) \circ \alpha = id_X$. For any $\beta \in \mathrm{Hom}_H(X, X)$, we have

$$(id_X \otimes \sigma)_*(\alpha \circ \beta) = \beta.$$

It yields that the map $(id_X \otimes \sigma)_*$ is surjective.

(2) $\Leftrightarrow$ (5). Applying the commutative diagram (1.7), we have the following commutative diagram:

$$\begin{array}{ccc}
\mathrm{Hom}_H(X^* \otimes X, E) & \xrightarrow{\sigma_*} & \mathrm{Hom}_H(X^* \otimes X, \Bbbk) \\
\Big\downarrow{\scriptstyle \Phi_{X^*,X,E}} & & \Big\downarrow{\scriptstyle \Phi_{X^*,X,\Bbbk}} \\
\mathrm{Hom}_H(X^*, E \otimes X^*) & \xrightarrow{(\sigma \otimes id_{X^*})_*} & \mathrm{Hom}_H(X^*, X^*)
\end{array}$$

Thus, σ_* is surjective if and only if $(\sigma \otimes id_{X^*})_*$ is surjective. If $(\sigma \otimes id_{X^*})_*$ is surjective, for the identity map id_{X^*}, there is a $\alpha \in \mathrm{Hom}_H(X^*, E \otimes X^*)$ such that

$$id_{X^*} = (\sigma \otimes id_{X^*})_*(\alpha) = (\sigma \otimes id_{X^*}) \circ \alpha.$$

This implies that $\sigma \otimes id_{X^*}$ is a split epimorphism. Conversely, if $\sigma \otimes id_{X^*}$ is a split epimorphism, there is a $\alpha \in \mathrm{Hom}_H(X^*, E \otimes X^*)$ such that

$$(\sigma \otimes id_{X^*}) \circ \alpha = id_{X^*}.$$

For any $\beta \in \mathrm{Hom}_H(X^*, X^*)$, we obtain that

$$(\sigma \otimes id_{X^*})_*(\alpha \circ \beta) = \beta.$$

It follows that the map $(\sigma \otimes id_{X^*})_*$ is surjective. □

Similarly, there are some equivalent conditions for $\Bbbk \nmid X \otimes X^*$. However, we only need the following characterization, which is useful in the study of the Green ring of H.

Proposition 1.2.11 *Let X be an indecomposable H-module. The following statements are equivalent*:

(1) $\Bbbk \nmid X \otimes X^*$.

(2) *The map* $E \otimes X \xrightarrow{\sigma \otimes id_X} X$ *is a split epimorphism.*

Proof Let Y be indecomposable such that $Y^* \cong X$ (such a Y exists as the order of S^2 is finite). Then $\Bbbk \nmid X \otimes X^*$ if and only if $\Bbbk \nmid (Y^* \otimes Y)^*$ if and only if $\Bbbk \nmid Y^* \otimes Y$. By Proposition 1.2.10, this is precisely that $\sigma \otimes id_{Y^*}$ is a split epimorphism, as desired. □

We have characterized $\Bbbk \nmid X^* \otimes X$ and $\Bbbk \nmid X \otimes X^*$ respectively in the previous two propositions. The following characterizations of $\Bbbk \mid X^* \otimes X$ and $\Bbbk \mid X \otimes X^*$ are useful.

Proposition 1.2.12 *Let X be an indecomposable H-module. The following statements are equivalent:*

(1) $\Bbbk \mid X^* \otimes X$.

(2) *The map* $X \otimes E \xrightarrow{id_X \otimes \sigma} X$ *is right almost split.*

Proof If $id_X \otimes \sigma$ is right almost split, it is not a split epimorphism. By Proposition 1.2.10, we have $\Bbbk \mid X^* \otimes X$. Conversely, if $\Bbbk \mid X^* \otimes X$, by Proposition 1.2.10, the morphism $id_X \otimes \sigma$ is not a split epimorphism. For any $\alpha \in \mathrm{Hom}_H(Y, X)$ which is not split epimorphism, the morphism $\Psi_{Y,X,\Bbbk}(\alpha) \in \mathrm{Hom}_H(X^* \otimes Y, \Bbbk)$ is also not split epimorphism by Proposition 1.2.5 (2). For the morphism $\Psi_{Y,X,\Bbbk}(\alpha)$, there is a morphism $\beta \in \mathrm{Hom}_H(X^* \otimes Y, E)$ such that

$$\sigma \circ \beta = \Psi_{Y,X,\Bbbk}(\alpha),$$

since σ is right almost split. Note that $\Psi^{-1}_{Y,X,E}(\beta) \in \mathrm{Hom}_H(Y, X \otimes E)$. We claim that the morphism $\Psi^{-1}_{Y,X,E}(\beta)$ satisfies the relation $(id_X \otimes \sigma) \circ \Psi^{-1}_{Y,X,E}(\beta) = \alpha$, and hence $id_X \otimes \sigma$ is right almost split. In fact, the commutative diagram (1.6) states that

$$\Psi_{Y,X,\Bbbk} \circ (id_X \otimes \sigma)_* = \sigma_* \circ \Psi_{Y,X,E}.$$

It follows that

$$\begin{aligned}\alpha &= \Psi^{-1}_{Y,X,\Bbbk}(\sigma \circ \beta) = (\Psi^{-1}_{Y,X,\Bbbk} \circ \sigma_*)(\beta)\\ &= ((id_X \otimes \sigma)_* \circ \Psi^{-1}_{Y,X,E})(\beta) = (id_X \otimes \sigma) \circ \Psi^{-1}_{Y,X,E}(\beta).\end{aligned}$$

This completes the proof. □

Similarly, we have the following result:

Proposition 1.2.13 *Let X be an indecomposable H-module. The following statements are equivalent:*

(1) $\Bbbk \mid X \otimes X^*$.

(2) *The map* $E \otimes X \xrightarrow{\sigma \otimes id_X} X$ *is right almost split.*

Remark 1.2.14 *An indecomposable module satisfying one of the equivalent conditions in Proposition* 1.2.12 *or in Proposition* 1.2.13 *is called a splitting trace module, see e.g.*, [26, 35, 95].

1.3 Bilinear forms on Green rings

In this section, we follow the approach of [8] and impose three bilinear forms on the Green ring of a finite dimensional Hopf algebra H. We will reveal the relationship among these bilinear forms.

Let $F(H)$ be the free abelian group generated by isomorphism classes $[X]$ of all finite dimensional H-modules X. The group $F(H)$ is in fact a ring with a multiplication given by the tensor product $[X][Y]=[X\otimes Y]$. The *Green ring* $r(H)$ of H is defined to be the quotient ring of $F(H)$ modulo the relations $[X\oplus Y]=[X]+[Y]$, for all H-modules X and Y. The identity of the associative ring $r(H)$ is represented by the trivial module $[\Bbbk]$. The set $\text{ind}(H)$ consisting of isomorphism classes of all indecomposable H-modules forms a $\mathbb{Z}$-basis of $r(H)$, see e.g., [19,40,47,82].

The *Grothendieck ring* $G_0(H)$ of H is the quotient ring of $F(H)$ modulo all short exact sequences of H-modules, i.e., $[Y]=[X]+[Z]$ if $0\to X\to Y\to Z\to 0$ is exact. The Grothendieck ring $G_0(H)$ possesses a $\mathbb{Z}$-basis given by isomorphism classes of simple H-modules. Both $r(H)$ and $G_0(H)$ are augmented $\mathbb{Z}$-algebras with the dimension augmentation. There is a natural ring epimorphism from $r(H)$ to $G_0(H)$ given by

$$\varphi: r(H)\to G_0(H),\quad [M]\mapsto \sum_{[V]}[M:V][V], \tag{1.8}$$

where $[M:V]$ is the multiplicity of V in the composition series of M and the sum $\sum_{[V]}$ runs over all non-isomorphic simple H-modules. If H is semisimple, the map φ is the identity map.

Let Z be an indecomposable H-module. If Z is non-projective, there is a unique almost split sequence $0\to X\to Y\to Z\to 0$ ending at Z. We follow the notation given in [4, Section 4, ChVI] and denote by $\delta_{[Z]}$ the element $[X]-[Y]+[Z]$ in $r(H)$. In case Z is projective, we define $\delta_{[Z]}:=[Z]-[\text{rad}Z]$. The following proposition gives a weaker condition for $\delta_{[Z]}=[X]-[Y]+[Z]$ in $r(H)$.

Proposition 1.3.1 *Let Z be an indecomposable non-projective H-module. If $0\to X\to Y\xrightarrow{\alpha} Z\to 0$ is a short exact sequence and the morphism α is right almost split, we still have*

$$\delta_{[Z]}=[X]-[Y]+[Z].$$

Proof Since the sequence

$$0\to X\to Y\xrightarrow{\alpha} Z\to 0 \tag{1.9}$$

is exact and the morphism α is right almost split, it follows from [4, Theorem 2.2, ChI] that the middle term Y has a decomposition $Y=Y_1\oplus Y_2$ such that the restriction of α to the summand Y_1, denoted by $\alpha|_{Y_1}$, is right minimal, and the restriction of α to the summand Y_2 is zero. We obtain that $\alpha|_{Y_1}$ is both right minimal and right almost split. By [4, Proposition 1.12, ChV], the sequence

$$0\to \ker(\alpha|_{Y_1})\xrightarrow{\iota} Y_1\xrightarrow{\alpha|_{Y_1}} Z\to 0$$

is almost split, where ι is the inclusion map. Thus,

$$\delta_{[Z]}=[\ker(\alpha|_{Y_1})]-[Y_1]+[Z].$$

Meanwhile, it is easy to see that the sequence

$$0 \to \ker(\alpha|_{Y_1}) \oplus Y_2 \xrightarrow{\iota \coprod id_{Y_2}} Y_1 \oplus Y_2 \xrightarrow{\alpha} Z \to 0 \tag{1.10}$$

is exact. Applying the short five lemma to the sequences (1.9) and (1.10), we obtain that

$$X \cong \ker(\alpha|_{Y_1}) \oplus Y_2.$$

In this case,

$$\begin{aligned}\delta_{[Z]} &= [\ker(\alpha|_{Y_1})] - [Y_1] + [Z]\\ &= [\ker(\alpha|_{Y_1}) \oplus Y_2] - [Y_1 \oplus Y_2] + [Z]\\ &= [X] - [Y] + [Z],\end{aligned}$$

as desired. □

For two H-modules X and Y, following from [8, 65, 94] we define

$$\langle [X], [Y] \rangle_1 := \dim_{\Bbbk} \mathrm{Hom}_H(X, Y).$$

Then, $\langle -, - \rangle_1$ extends to a $\mathbb{Z}$-bilinear form on $r(H)$. The following lemma follows from Proposition 4.1, Theorem 4.3 and Theorem 4.4 in [4, ChVI], so we omit its proof.

Lemma 1.3.2 *The following results hold in* $r(H)$:

(1) *For two indecomposable modules* X *and* Z, *we have* $\langle [X], \delta_{[Z]} \rangle_1 = \delta_{[X],[Z]}$, *where* $\delta_{[X],[Z]}$ *is equal to* 1 *if* $X \cong Z$, *and* 0 *otherwise.*

(2) *For any* $x \in r(H)$, *we have* $x = \sum_{[M] \in \mathrm{ind}(H)} \langle x, \delta_{[M]} \rangle_1 [M]$.

(3) *The set* $\{\delta_{[M]} \mid [M] \in \mathrm{ind}(H)\}$ *is linearly independent in* $r(H)$.

(4) H *is of finite representation type if and only if* $\{\delta_{[M]} \mid [M] \in \mathrm{ind}(H)\}$ *forms a* $\mathbb{Z}$-*basis of* $r(H)$.

(5) H *is of finite representation type if and only if* $\{\delta_{[M]} \mid [M] \in \mathrm{ind}(H)$ *and* M *not projective*$\}$ *forms a* $\mathbb{Z}$-*basis of* $\ker \varphi$, *where* φ *is the map given in* (1.8).

Remark 1.3.3 *It follows from Lemma* 1.3.2 (2) *that the form* $\langle -, - \rangle_1$ *is non-degenerate in the sense that given* $0 \neq x \in r(H)$, *there is* $y \in r(H)$ *such that* $\langle x, y \rangle_1 \neq 0$. *If* H *is of finite representation type, it can be seen from Lemma* 1.3.2 *that the set* $\{[M], \delta_{[M]} \mid [M] \in \mathrm{ind}(H)\}$ *forms a pair of dual bases of* $r(H)$ *with respect to the form* $\langle -, - \rangle_1$. *In this case, any* x *in* $r(H)$ *can be written as follows:*

$$x = \sum_{[M] \in \mathrm{ind}(H)} \langle [M], x \rangle_1 \delta_{[M]}.$$

We use the non-degeneracy of the form $\langle -, - \rangle_1$ to give an equivalent condition for H to be of finite representation type.

Proposition 1.3.4 *The finite dimensional Hopf algebra H is of finite representation type if and only if for any indecomposable module X, there are only finitely many indecomposable modules M such that* $\mathrm{Hom}_H(M,X)\neq 0$.

Proof For any indecomposable module X, if there are only finitely many indecomposable modules M such that $\mathrm{Hom}_H(M,X)\neq 0$, then the sum

$$\sum_{[M]\in\mathrm{ind}(H)} \dim_{\Bbbk}\mathrm{Hom}_H(M,X)\delta_{[M]}$$

is a finite sum. We have

$$\begin{aligned}&\langle [M],[X]-\sum_{[M]\in\mathrm{ind}(H)} \dim_{\Bbbk}\mathrm{Hom}_H(M,X)\delta_{[M]}\rangle_1\\ =\ &\langle [M],[X]\rangle_1-\dim_{\Bbbk}\mathrm{Hom}_H(M,X)=0.\end{aligned}$$

This implies that

$$[X]=\sum_{[M]\in\mathrm{ind}(H)} \dim_{\Bbbk}\mathrm{Hom}_H(M,X)\delta_{[M]}$$

by the non-degeneracy of the form $\langle -,-\rangle_1$. Thus, $\{\delta_{[M]}\mid [M]\in\mathrm{ind}(H)\}$ is a basis of $r(H)$, and hence H is of finite representation type by Lemma 1.3.2 (4). □

Let $\mathcal{P}(X,Y)$ be the space of morphisms from X to Y which factor through a projective module. By a similar way to [8], we define another bilinear form on $r(H)$ as follows:

$$\langle [X],[Y]\rangle_2:=\dim_{\Bbbk}\mathcal{P}(X,Y).$$

Let $*$ denote the duality operator of $r(H)$ induced by the duality functor: $[X]^*=[X^*]$. Then $*$ is an anti-automorphism of $r(H)$. Obviously, if S^2 is inner, then $*$ is an involution [51]. The forms $\langle -,-\rangle_1$ and $\langle -,-\rangle_2$ have the following properties:

Proposition 1.3.5 *Let X, Y and Z be H-modules.*

(1) $\langle [X][Y],[Z]\rangle_1=\langle [X],[Z][Y]^*\rangle_1$ *and* $\langle [X],[Y][Z]\rangle_1=\langle [Y]^*[X],[Z]\rangle_1$.

(2) $\langle [X][Y],[Z]\rangle_2=\langle [X],[Z][Y]^*\rangle_2$ *and* $\langle [X],[Y][Z]\rangle_2=\langle [Y]^*[X],[Z]\rangle_2$.

Proof (1) It follows from Lemma 1.2.4.

(2) If $\alpha\in\mathrm{Hom}_H(X\otimes Y,Z)$ factors through a projective module P, then $\Phi_{X,Y,Z}(\alpha)$ factors through the projective module $P\otimes Y^*$ by Lemma 1.2.4 (1). Thus,

$$\Phi_{X,Y,Z}(\mathcal{P}(X\otimes Y,Z))\subseteq\mathcal{P}(X,Z\otimes Y^*).$$

Conversely, for any $\beta\in\mathcal{P}(X,Z\otimes Y^*)$ which factors through a projective module P, by Lemma 1.2.4 (1), the morphism $\Phi^{-1}_{X,Y,Z}(\beta)$ factors through the projective module $P\otimes Y$. We obtain that

$$\Phi_{X,Y,Z}(\mathcal{P}(X\otimes Y,Z))=\mathcal{P}(X,Z\otimes Y^*).$$

Similarly, we have

$$\Psi_{X,Y,Z}(\mathcal{P}(X, Y\otimes Z)) = \mathcal{P}(Y^*\otimes X, Z).$$

We are done. □

Let Ω and Ω^{-1} denote the *syzygy functor* and *cosyzygy functor* of H-mod respectively. Namely, ΩM is the kernel of the projective cover $P_M \to M$, and $\Omega^{-1}M$ is the cokernel of the injective envelope $M \to I_M$. Denote by $\delta^*_{[M]}$ the image of $\delta_{[M]}$ under the duality operator $*$ of $r(H)$. The following lemma is a generalization of [8, Proposition 2.1] to the case of the Green ring $r(H)$. We omit the proof since it is similar to the proof of [8, Proposition 2.1].

Lemma 1.3.6 *Let M be an indecomposable H-module and $P_{\Bbbk}$ the projective cover of the trivial module $\Bbbk$. The following results hold in $r(H)$:*

(1) $([I_M]-[\Omega^{-1}M])\delta_{[P_{\Bbbk}]} = \delta_{[P_{\Bbbk}]}([I_M]-[\Omega^{-1}M]) = [M]$ and $([P_M]-[\Omega M])\delta^*_{[P_{\Bbbk}]} = \delta^*_{[P_{\Bbbk}]}([P_M]-[\Omega M]) = [M]$. Moreover, $\delta_{[P_{\Bbbk}]}\delta^*_{[P_{\Bbbk}]} = \delta^*_{[P_{\Bbbk}]}\delta_{[P_{\Bbbk}]} = 1$.

(2) $[M]\delta_{[P_{\Bbbk}]} = \delta_{[P_{\Bbbk}]}[M] = [P_M]-[\Omega M]$ and $[M]\delta^*_{[P_{\Bbbk}]} = \delta^*_{[P_{\Bbbk}]}[M] = [I_M]-[\Omega^{-1}M]$. Thus, $\delta_{[P_{\Bbbk}]}$ and $\delta^*_{[P_{\Bbbk}]}$ are both central units of $r(H)$.

The following result explores a relationship between the forms $\langle -,-\rangle_1$ and $\langle -,-\rangle_2$. We refer to [8, Corollary 2.3] for a similar result for the Green ring of a finite group.

Proposition 1.3.7 *Let X and Y be two H-modules.*

(1) $\langle [X],[Y]\rangle_2$ is equal to the multiplicity of $P_{\Bbbk}$ in the direct sum decomposition of $Y^*\otimes X$.

(2) $\langle [X],[Y]\rangle_2 = \langle [X],[Y]\delta_{[P_{\Bbbk}]}\rangle_1 = \langle [X]\delta^*_{[P_{\Bbbk}]},[Y]\rangle_1$.

(3) $\langle [X],[Y]\rangle_1 = \langle [X]\delta_{[P_{\Bbbk}]},[Y]\rangle_2 = \langle [X],[Y]\delta^*_{[P_{\Bbbk}]}\rangle_2$.

Proof (1) For any non-zero morphism $\alpha\in\mathcal{P}(Y^*\otimes X,\Bbbk)$, if α factors through an indecomposable projective module P, then $\alpha=\beta\circ\gamma$ for some $\beta: P\to\Bbbk$ and $\gamma: Y^*\otimes X\to P$. Since β is surjective, P is the projective cover of $\Bbbk$ and hence $P\cong P_{\Bbbk}$. Note that $\mathrm{rad}P_{\Bbbk}$ is the unique maximal submodule of $P_{\Bbbk}$. The image of the morphism γ is either contained in $\mathrm{rad}P_{\Bbbk}$ or equal to $P_{\Bbbk}$. For the former case, $\alpha=\beta\circ\gamma=0$, a contradiction. Thus, the latter case holds and the morphism γ is surjective, and hence $P_{\Bbbk}$ is a direct summand of $Y^*\otimes X$. Now, if α factors through a projective module P and $P\cong\bigoplus_i P_i$ for some indecomposable projective modules P_i. Then $\alpha=\sum_i\beta_i\circ\gamma_i$ for some $\beta_i: P_i\to\Bbbk$ and $\gamma_i: Y^*\otimes X\to P_i$. We have proved that $\beta_i\circ\gamma_i\neq 0$ if and only if $P_i\cong P_{\Bbbk}$. It follows that $\dim_{\Bbbk}\mathcal{P}(Y^*\otimes X,\Bbbk)$ is equal to the multiplicity of $P_{\Bbbk}$ in the direct sum decomposition of $Y^*\otimes X$, while $\dim_{\Bbbk}\mathcal{P}(Y^*\otimes X,\Bbbk)$ is equal to $\dim_{\Bbbk}\mathcal{P}(X,Y)$ by Proposition 1.3.5 (2), as desired.

(2) It follows from Part (1) that

$$\langle [X],[Y]\rangle_2 = \langle [Y]^*[X],\delta_{[P_{\Bbbk}]}\rangle_1.$$

By Proposition 1.3.5, we have

$$\begin{aligned}\langle [Y]^*[X], \delta_{[P_k]}\rangle_1 &= \langle [X], [Y]\delta_{[P_k]}\rangle_1 = \langle [X]\delta^*_{[P_k]}\delta_{[P_k]}, [Y]\delta_{[P_k]}\rangle_1 \\ &= \langle [X]\delta^*_{[P_k]}, [Y]\delta_{[P_k]}\delta^*_{[P_k]}\rangle_1 = \langle [X]\delta^*_{[P_k]}, [Y]\rangle_1.\end{aligned}$$

(3) It follows from Part (2) and the fact that

$$\delta_{[P_k]}\delta^*_{[P_k]} = \delta^*_{[P_k]}\delta_{[P_k]} = 1.$$

□

Corollary 1.3.8 *Let X be an indecomposable H-module and V a simple H-module. Then $\langle [X], [V]\rangle_2 = \delta_{[X],[P_V]}$.*

Proof It follows from Proposition 1.3.7 that

$$\langle [X], [V]\rangle_2 = \langle [X], [V]\delta_{[P_k]}\rangle_1.$$

By Lemma 1.3.6,

$$\langle [X], [V]\delta_{[P_k]}\rangle_1 = \langle [X], [P_V]\rangle_1 - \langle [X], [\Omega V]\rangle_1,$$

which is equal to 1 if $X \cong P_V$, and 0 otherwise. □

Remark 1.3.9 *Let H be of finite representation type. It follows from Proposition 1.3.7 (3) that the set $\{[M]\delta_{[P_k]}, \delta_{[M]} \mid [M] \in \mathrm{ind}(H)\}$ or $\{[M], \delta_{[M]}\delta^*_{[P_k]} \mid [M] \in \mathrm{ind}(H)\}$ forms a pair of dual bases of $r(H)$ with respect to the form $\langle -,-\rangle_2$. Hence the form $\langle -,-\rangle_2$ is the same as $\langle -,-\rangle_1$ up to a unit. Namely, we have*

$$\langle -,-\rangle_1 = \langle -\delta_{[P_k]}, -\rangle_2 = \langle -, -\delta^*_{[P_k]}\rangle_2.$$

For two H-modules X and Y, we define

$$\langle [X], [Y]\rangle_3 := \langle [X], [Y]\rangle_1 - \langle [X], [Y]\rangle_2.$$

It follows from Proposition 1.3.7 that

$$\langle [X], [Y]\rangle_3 = \langle [X], [Y](1 - \delta_{[P_k]})\rangle_1 = \langle [X](1 - \delta^*_{[P_k]}), [Y]\rangle_1.$$

Moreover, we have the following result which gives another relationship between the forms $\langle -,-\rangle_3$ and $\langle -,-\rangle_1$.

Proposition 1.3.10 *Let X and Y be two H-modules.*

(1) *If X is projective, then $\langle [X], [Y]\rangle_3 = 0$.*

(2) *If X is indecomposable and non-projective, then*

$$\langle [X], [Y]\rangle_3 = \langle [X], [Y]\rangle_1 + \langle [\Omega^{-1}X], [Y]\rangle_1 - \sum_{[V]} [Y : V]\langle [\Omega^{-1}X], [V]\rangle_1,$$

where the sum $\sum_{[V]}$ runs over all non-isomorphic simple H-modules and $[Y : V]$ is the multiplicity of V in the composition series of Y. In particular, $\langle [X], [Y]\rangle_3 = \langle [X], [Y]\rangle_1$ if Y is simple.

Proof (1) It follows from the fact that $\mathcal{P}(X,Y) = \mathrm{Hom}_H(X,Y)$ if X is projective.

(2) For any simple H-module V, on the one hand, $\langle [X],[V]\rangle_2 = 0$ by Corollary 1.3.8. On the other hand,

$$\langle [X],[V]\rangle_2 = \langle [X]\delta^*_{[P_{\Bbbk}]},[V]\rangle_1,$$

which is equal to $\langle [I_X]-[\Omega^{-1}X],[V]\rangle_1$ by Lemma 1.3.6. It follows that

$$\langle [I_X],[V]\rangle_1 = \langle [\Omega^{-1}X],[V]\rangle_1.$$

Now

$$\begin{aligned}\langle [X],[Y]\rangle_3 &= \langle [X](1-\delta^*_{[P_{\Bbbk}]}),[Y]\rangle_1\\ &= \langle [X],[Y]\rangle_1 + \langle [\Omega^{-1}X],[Y]\rangle_1 - \langle [I_X],[Y]\rangle_1\\ &= \langle [X],[Y]\rangle_1 + \langle [\Omega^{-1}X],[Y]\rangle_1 - \sum_{[V]}[Y:V]\langle [I_X],[V]\rangle_1\\ &= \langle [X],[Y]\rangle_1 + \langle [\Omega^{-1}X],[Y]\rangle_1 - \sum_{[V]}[Y:V]\langle [\Omega^{-1}X],[V]\rangle_1,\end{aligned}$$

as desired. □

The left radical of the form $\langle -,-\rangle_3$ is the set

$$\{x\in r(H) \mid \langle x,y\rangle_3 = 0,\quad \text{for all } y\in r(H)\}.$$

This set is exactly the set $\{x\in r(H)\mid x(1-\delta^*_{[P_{\Bbbk}]})=0\}$. Similarly, the right radical of the form $\langle -,-\rangle_3$ is exactly the set $\{x\in r(H)\mid x(1-\delta_{[P_{\Bbbk}]})=0\}$. The left and right radicals of the form coincide since

$$\delta_{[P_{\Bbbk}]}\delta^*_{[P_{\Bbbk}]} = \delta^*_{[P_{\Bbbk}]}\delta_{[P_{\Bbbk}]} = 1.$$

Note that $[P](1-\delta_{[P_{\Bbbk}]})=0$ for any projective module P. Thus, the projective ideal $\mathcal{P}$ of $r(H)$ generated by isomorphism classes of projective H-modules is contained in the radical of the form. For further results about the radical of the form, we need the following lemma:

Lemma 1.3.11 *Let M and Z be two indecomposable H-modules.*

(1) $\langle [M],\delta_{[Z]}\rangle_3 = \delta_{[M],[Z]} + \delta_{[\Omega^{-1}M],[Z]} - \delta_{[I_M],[Z]}$.

(2) $\delta_{[M]} = \begin{cases} -\delta_{[\Omega^{-1}M]}\delta_{[P_{\Bbbk}]}, & M \text{ is not projective},\\ [\mathrm{top}M]\delta_{[P_{\Bbbk}]}, & M \text{ is projective}.\end{cases}$

Proof (1) Note that

$$\langle [M],\delta_{[Z]}\rangle_3 = \langle [M](1-\delta^*_{[P_{\Bbbk}]}),\delta_{[Z]}\rangle_1.$$

It follows that

$$\begin{aligned}\langle [M](1-\delta^*_{[P_{\Bbbk}]}), \delta_{[Z]}\rangle_1 &= \langle [M], \delta_{[Z]}\rangle_1 - \langle [I_M]-[\Omega^{-1}M], \delta_{[Z]}\rangle_1 \\ &= \langle [M], \delta_{[Z]}\rangle_1 + \langle [\Omega^{-1}M], \delta_{[Z]}\rangle_1 - \langle [I_M], \delta_{[Z]}\rangle_1 \\ &= \delta_{[M],[Z]} + \delta_{[\Omega^{-1}M],[Z]} - \delta_{[I_M],[Z]}.\end{aligned}$$

(2) Suppose M is not projective. For any indecomposable module X, we have

$$\begin{aligned}&\langle [X], \delta_{[M]} + \delta_{[\Omega^{-1}M]}\delta_{[P_{\Bbbk}]}\rangle_1 \\ &= \langle [X], \delta_{[M]}\rangle_1 + \langle [X]\delta^*_{[P_{\Bbbk}]}, \delta_{[\Omega^{-1}M]}\rangle_1 \quad \text{(by Proposition 1.3.7(2))} \\ &= \langle [X], \delta_{[M]}\rangle_1 + \langle [I_X]-[\Omega^{-1}X], \delta_{[\Omega^{-1}M]}\rangle_1 \\ &= \delta_{[X],[M]} - \delta_{[\Omega^{-1}X],[\Omega^{-1}M]} + \delta_{[I_X],[\Omega^{-1}M]} \\ &= 0.\end{aligned}$$

Thus,

$$\delta_{[M]} = -\delta_{[\Omega^{-1}M]}\delta_{[P_{\Bbbk}]}$$

since the form $\langle -,-\rangle_1$ is non-degenerate. Now suppose M is projective, for any indecomposable module X, we have

$$\begin{aligned}\langle [X], \delta_{[M]} - [\mathrm{top}M]\delta_{[P_{\Bbbk}]}\rangle_1 &= \langle [X], \delta_{[M]}\rangle_1 - \langle [X], [\mathrm{top}M]\delta_{[P_{\Bbbk}]}\rangle_1 \\ &= \langle [X], \delta_{[M]}\rangle_1 - \langle [X], [\mathrm{top}M]\rangle_2 \\ &= \delta_{[X],[M]} - \delta_{[X],[M]} \quad \text{(by Corollary 1.3.8)} \\ &= 0.\end{aligned}$$

Thus, $\delta_{[M]} = [\mathrm{top}M]\delta_{[P_{\Bbbk}]}$. □

Recall that an H-module M is called a *periodic module* of period n if $\Omega^n M \cong M$ for a minimal natural number n (see e.g., [12]).

Theorem 1.3.12 *Let H be of finite representation type. The radical of the form $\langle -,-\rangle_3$ is equal to $\mathcal{P}$ if and only if there are no periodic modules of even period.*

Proof Note that the projective ideal $\mathcal{P}$ of $r(H)$ is contained in the radical of the form $\langle -,-\rangle_3$. If $\mathcal{P}$ is properly contained in the radical of the form $\langle -,-\rangle_3$, there exist some indecomposable non-projective H-modules M such that $\sum_{[M]}\lambda_{[M]}[M]$ is a non-zero element in the radical of the form. For any indecomposable non-projective module Z, by Lemma 1.3.11 (1), we have

$$0 = \langle \sum_{[M]} \lambda_{[M]}[M], \delta_{[\Omega^i Z]}\rangle_3 = \lambda_{[\Omega^i Z]} + \lambda_{[\Omega^{i+1}Z]}, \quad \text{for } i \geqslant 0.$$

It follows that

$$\lambda_{[\Omega^i Z]} = (-1)^i \lambda_{[Z]}, \quad \text{for } i \geqslant 0.$$

This forces $\lambda_{[Z]} = 0$ if Z is a periodic module of odd period. However, $\sum_{[M]} \lambda_{[M]}[M]$ is not zero, which implies that there exists a periodic module M of even period with $\lambda_{[M]} \neq 0$. Conversely, suppose the radical of the form $\langle -, - \rangle_3$ is equal to $\mathcal{P}$. We claim that the category H-mod has no periodic modules of even period. Otherwise, if M is a periodic module of even period $2s$. It follows from Lemma 1.3.11 (2) that

$$\sum_{i=1}^{2s}(-1)^i \delta_{[\Omega^i M]}(1 - \delta^*_{[P_\Bbbk]}) = \sum_{i=1}^{2s}(-1)^i(\delta_{[\Omega^i M]} + \delta_{[\Omega^{i-1} M]}) = 0.$$

Thus, $\sum_{i=1}^{2s}(-1)^i \delta_{[\Omega^i M]}$ belongs to the radical of the form $\langle -, - \rangle_3$. Suppose

$$\sum_{i=1}^{2s}(-1)^i \delta_{[\Omega^i M]} = \sum_j \lambda_j [P_j],$$

for some indecomposable projective modules P_j. By Remark 1.3.3, $[P_j]$ can be written as

$$[P_j] = \sum_{[M] \in \operatorname{ind}(H)} \langle [M], [P_j] \rangle_1 \delta_{[M]}.$$

It follows that

$$\sum_{i=1}^{2s}(-1)^i \delta_{[\Omega^i M]} = \sum_j \sum_{[M] \in \operatorname{ind}(H)} \lambda_j \langle [M], [P_j] \rangle_1 \delta_{[M]}.$$

Comparing the coefficient of $\delta_{[\Omega^i M]}$ in both two sides of the above equality, we obtain that

$$(-1)^i = \sum_j \lambda_j \dim_\Bbbk \operatorname{Hom}_H(\Omega^i M, P_j) = \sum_j \lambda_j \dim_\Bbbk \operatorname{Ext}^i_H(M, P_j) = 0,$$

a contradiction. □

1.4 Some ring-theoretical properties

In this section, we use a bilinear form $(-, -)$ to give several one-sided ideals of $r(H)$ and these ideals provide a little more information about the Jacobson radical and central primitive idempotents of $r(H)$. It is known that the Grothendieck ring $G_0(H)$ of H is a quotient ring of $r(H)$. We describe this quotient ring clearly:

$$r(H)/\mathcal{P}^\perp \cong \mathrm{G}_0(H),$$

where $\mathcal{P}^{\perp}$ is orthogonal to the projective ideal $\mathcal{P}$ with respect to the form $(-,-)$. This isomorphism will be used later to characterize when the Jacobson radical of $r(H)$ is equal to the intersection $\mathcal{P}\cap\mathcal{P}^{\perp}$.

Note that the $\mathbb{Z}$-bilinear form $\langle -,-\rangle_1$ is not associative in general. However, we may modify it as follows:

$$([X],[Y]) := \langle [X],[Y]^*\rangle_1 = \dim_{\Bbbk} \mathrm{Hom}_H(X, Y^*). \tag{1.11}$$

Then $(-,-)$ extends to a $\mathbb{Z}$-bilinear form on $r(H)$.

Lemma 1.4.1 *For H-modules X, Y and Z, the form $(-,-)$ satisfies the following properties:*

(1) $([X][Y],[Z]) = ([X],[Y][Z])$.

(2) $([X],[Y]) = ([Y]^{**},[X])$. *If S^2 is inner, then* $([X],[Y]) = ([Y],[X])$.

Proof (1) The associativity of the form follows from Lemma 1.2.4 (1), i.e.,

$$\begin{aligned}([X][Y],[Z]) &= \dim_{\Bbbk} \mathrm{Hom}_H(X\otimes Y, Z^*)\\ &= \dim_{\Bbbk} \mathrm{Hom}_H(X, (Y\otimes Z)^*)\\ &= ([X],[Y][Z]).\end{aligned}$$

(2) The $\Bbbk$-linear isomorphism $\mathrm{Hom}_H(X,Y^*)\cong \mathrm{Hom}_H(Y^{**},X^*)$ following from Lemma 1.2.4 (see also [51]) implies that $([X],[Y]) = ([Y]^{**},[X])$. If S^2 is inner, the anti-automorphism $*$ of $r(H)$ is an involution. In this case, $([X],[Y]) = ([Y],[X])$. □

The following result can be deduced directly from Lemma 1.3.2.

Lemma 1.4.2 *The following results hold in $r(H)$:*

(1) *For two indecomposable modules X and Z, we have*

$$(\delta^*_{[Z]},[X]) = \delta_{[Z],[X]}.$$

(2) *For any $x\in r(H)$, we have*

$$x = \sum_{[M]\in \mathrm{ind}(H)} (\delta^*_{[M]}, x)[M].$$

(3) *The form $(-,-)$ is non-degenerate.*

As a consequence, we obtain the following Frobenius property of $r(H)$:

Proposition 1.4.3 *Let H be of finite representation type. The Green ring $r(H)$ is a Frobenius $\mathbb{Z}$-algebra. Moreover, $r(H)$ is a symmetric $\mathbb{Z}$-algebra if S^2 is inner.*

Proof Note that

$$r(H)^{\vee} := \mathrm{Hom}_{\mathbb{Z}}(r(H),\mathbb{Z})$$

is a $(r(H), r(H))$-bimodule via $(afb)(x) = f(bxa)$, for $a, b, x \in r(H)$ and $f \in r(H)^\vee$. Since H is of finite representation type, the form $(-,-)$ is associative and non-degenerate with a pair of dual bases $\{\delta^*_{[M]}, [M] \mid [M] \in \operatorname{ind}(H)\}$. Thus, the map ρ from $r(H)$ to $r(H)^\vee$ given by $x \mapsto (-, x)$ is a left $r(H)$-module isomorphism, and hence $r(H)$ is a Frobenius $\mathbb{Z}$-algebra. Moreover, if the square of the antipode of H is inner, the bilinear form $(-,-)$ is symmetric and hence ρ is a $(r(H), r(H))$-bimodule isomorphism. It follows that $r(H)$ is a symmetric $\mathbb{Z}$-algebra. □

Remark 1.4.4 *Let H be of finite representation type.*

(1) *The Green ring $r(H)$ is a Frobenius $\mathbb{Z}$-algebra with a pair of dual bases $\{\delta^*_{[M]}, [M] \mid [M] \in \operatorname{ind}(H)\}$ with respect to the form $(-,-)$. Lemma 1.4.2(2) is now equivalent to the following equality:*

$$x = \sum_{[M]\in \operatorname{ind}(H)} (x, [M])\delta^*_{[M]}, \quad \text{for } x \in r(H).$$

This means that the transformation matrix from the basis $\{\delta^*_{[M]} \mid [M] \in \operatorname{ind}(H)\}$ to the standard basis $\operatorname{ind}(H)$ is an invertible integer matrix with entries

$$([X], [Y]) = \dim_{\Bbbk} \operatorname{Hom}_H(X, Y^*), \quad \text{for } [X], [Y] \in \operatorname{ind}(H).$$

(2) *If H is semisimple, then S^2 is inner [46] and $\delta^*_{[M]} = [M]^* = [M^*]$. In this case, $r(H) = G_0(H)$ which is a symmetric algebra over $\mathbb{Z}$ with a pair of dual bases $\{[M^*], [M] \mid [M] \in \operatorname{ind}(H)\}$ (see Proposition 1.4.3). It is a semiprime ring in the sense that it has no nonzero nilpotent ideals [51]. We refer to [94] for more details in the semisimple case.*

The bilinear form $(-,-)$ can be used to describe the relationship between the Green ring $r(H)$ and the Grothendieck ring $G_0(H)$ of H. Let $\mathcal{P}^\perp$ be the subgroup of $r(H)$ which is orthogonal to $\mathcal{P}$ with respect to the form $(-,-)$. Then $\mathcal{P}^\perp$ is a two-sided ideal of $r(H)$.

Proposition 1.4.5 *The Grothendieck ring $G_0(H)$ is isomorphic to the quotient ring $r(H)/\mathcal{P}^\perp$.*

Proof Observe that the natural morphism φ given in (1.8) is surjective. It is sufficient to show that $\ker\varphi = \mathcal{P}^\perp$. Suppose

$$\sum_{[M]\in \operatorname{ind}(H)} \lambda_{[M]}[M] \in \ker\varphi,$$

where $\lambda_{[M]} \in \mathbb{Z}$. Then

$$\sum_{[V]} \sum_{[M]\in \operatorname{ind}(H)} \lambda_{[M]}[M:V][V] = 0.$$

Note that a short exact sequence tensoring over $\Bbbk$ with a projective module P is split. It follows that the equality $[M][P]=\sum_{[V]}[M:V][V][P]$ holds in $r(H)$, and hence

$$\begin{aligned}\left(\sum_{[M]\in \mathrm{ind}(H)} \lambda_{[M]}[M],[P]\right) &= \left(\sum_{[M]\in \mathrm{ind}(H)} \lambda_{[M]}[M][P],[\Bbbk]\right)\\ &= \left(\sum_{[V]}\sum_{[M]\in \mathrm{ind}(H)} \lambda_{[M]}[M:V][V][P],[\Bbbk]\right)\\ &= 0.\end{aligned}$$

This implies that

$$\sum_{[M]\in \mathrm{ind}(H)} \lambda_{[M]}[M]\in \mathcal{P}^{\perp}.$$

Now, we assume

$$\sum_{[M]\in \mathrm{ind}(H)} \lambda_{[M]}[M]\in \mathcal{P}^{\perp}.$$

Note that $[P]y\in\mathcal{P}$ for any $y\in r(H)$ and $[P]\in\mathcal{P}$. We have

$$\left(\sum_{[M]\in \mathrm{ind}(H)} \lambda_{[M]}[M][P],y\right) = \left(\sum_{[M]\in \mathrm{ind}(H)} \lambda_{[M]}[M],[P]y\right)=0.$$

This implies that

$$\sum_{[M]\in \mathrm{ind}(H)} \lambda_{[M]}[M][P]=0$$

as the form $(-,-)$ is non-degenerate. Replacing $[M][P]$ by $\sum_{[V]}[M:V][V][P]$, we obtain the following equality:

$$\sum_{[V]}\sum_{[M]\in \mathrm{ind}(H)} \lambda_{[M]}[M:V][V][P]=0. \tag{1.12}$$

Note that $\mathcal{P}$ is a $G_0(H)$-module under the action given by

$$[V][P]=[V\otimes P]\in\mathcal{P}.$$

Moreover, the $G_0(H)$-module $\mathcal{P}$ is faithful, see [51, Section 3.1]. It follows from (1.12) that

$$\sum_{[V]}\sum_{[M]\in \mathrm{ind}(H)} \lambda_{[M]}[M:V][V]=0, \quad \text{i.e.,} \quad \sum_{[M]\in \mathrm{ind}(H)} \lambda_{[M]}[M]\in\ker\varphi. \qquad \square$$

Now we turn to the special element $\delta_{[\Bbbk]}$, which plays an important role in the study of the Green ring $r(H)$. For any indecomposable module X, the elements $[X]$, $\delta_{[\Bbbk]}$ and $\delta_{[X]}$ satisfy the following relations:

Theorem 1.4.6 *Let X be an indecomposable H-module.*

(1) $\Bbbk \nmid X^* \otimes X$ *if and only if* $[X]\delta_{[\Bbbk]} = 0$.

(2) $\Bbbk \nmid X \otimes X^*$ *if and only if* $\delta_{[\Bbbk]}[X] = 0$.

(3) $\Bbbk \mid X^* \otimes X$ *if and only if* $[X]\delta_{[\Bbbk]} = \delta_{[X]}$.

(4) $\Bbbk \mid X \otimes X^*$ *if and only if* $\delta_{[\Bbbk]}[X] = \delta_{[X]}$.

Proof If H is semisimple, then $\Bbbk \mid X^* \otimes X$ and $\Bbbk \mid X \otimes X^*$. In this case, Part (3) and Part (4) hold obviously, because $\delta_{[\Bbbk]} = [\Bbbk]$ and $\delta_{[X]} = [X]$. Assume H is not semisimple, we only show Part (1) and Part (3) and the proofs of Part (2) and Part (4) are similar.

(1) If $\Bbbk \nmid X^* \otimes X$, by Proposition 1.2.10, the morphism $id_X \otimes \sigma$ is a split epimorphism. It follows from (1.5) that

$$[X \otimes E] = [X \otimes \tau(\Bbbk)] + [X],$$

and hence $[X]\delta_{[\Bbbk]} = 0$. Conversely, if $[X]\delta_{[\Bbbk]} = 0$, then

$$0 = (([X]\delta_{[\Bbbk]})^*, [X]) = (\delta^*_{[\Bbbk]}, [X]^*[X]).$$

This means that $\Bbbk \nmid X^* \otimes X$.

(3) If $\Bbbk \mid X^* \otimes X$, then the morphism $id_X \otimes \sigma$ is right almost split by Proposition 1.2.12. It follows from (1.5) and Proposition 1.3.1 that

$$\delta_{[X]} = [X \otimes \tau(\Bbbk)] - [X \otimes E] + [X] = [X]\delta_{[\Bbbk]}.$$

Conversely, if $[X]\delta_{[\Bbbk]} = \delta_{[X]}$, then

$$1 = (\delta^*_{[X]}, [X]) = (([X]\delta_{[\Bbbk]})^*, [X]) = (\delta^*_{[\Bbbk]}, [X]^*[X]).$$

It follows that $\Bbbk \mid X^* \otimes X$. □

As an application of Theorem 1.4.6, we are able to determine the multiplicity of the trivial module $\Bbbk$ in the decomposition of the tensor product $X \otimes X^*$ and $X^* \otimes X$ respectively. For the case where H is semisimple over a field $\Bbbk$ of characteristic 0, this was done by Zhu [99, Lemma 1], also see [94, Proposition 2.1].

Corollary 1.4.7 *Let X be an indecomposable H-module.*

(1) *The multiplicity of* $\Bbbk$ *in* $X^* \otimes X$ *is either* 0 *or* 1.

(2) *The multiplicity of* $\Bbbk$ *in* $X \otimes X^*$ *is either* 0 *or* 1.

Proof (1) We only prove Part (1), the proof of Part (2) is similar. Note that the multiplicity of $\Bbbk$ in $X^* \otimes X$ is $(\delta^*_{[\Bbbk]}, [X^*][X])$. By Theorem 1.4.6, we have

$$(\delta^*_{[\Bbbk]}, [X^*][X]) = (([X]\delta_{[\Bbbk]})^*, [X]) = \begin{cases} 0, & \Bbbk \nmid X^* \otimes X, \\ 1, & \Bbbk \mid X^* \otimes X, \end{cases}$$

as desired. □

The following result can be deduced from Theorem 1.4.6.

Proposition 1.4.8 *Let* $0 \to X \to Y \to Z \to 0$ *be an almost split sequence of* H*-modules.*

(1) $\Bbbk \mid Z \otimes Z^*$ *if and only if* $\Bbbk \mid X \otimes X^*$.

(2) $\Bbbk \mid Z^* \otimes Z$ *if and only if* $\Bbbk \mid X^* \otimes X$.

Proof Applying the duality functor $*$ to the almost split sequence $0 \to X \to Y \to Z \to 0$, we get the almost split sequence $0 \to Z^* \to Y^* \to X^* \to 0$, see [4, P144]. Note that both Z and X^* are indecomposable, see [4, Proposition 1.14, ChV]. This implies that

$$\delta_{[Z]}^* = \delta_{[X^*]}.$$

(1) If $\Bbbk \mid Z \otimes Z^*$, by Theorem 1.4.6, we have $\delta_{[\Bbbk]}[Z] = \delta_{[Z]}$. We claim that $\Bbbk \mid X \otimes X^*$. Otherwise, $\Bbbk \nmid X \otimes X^*$, and hence $\Bbbk \nmid X^{**} \otimes X^*$. This leads to $[X^*]\delta_{[\Bbbk]} = 0$ by Theorem 1.4.6. However,

$$1 = (\delta_{[X^*]}^*, [X^*]) = (\delta_{[Z]}^{**}, [X^*]) = ([X^*], \delta_{[Z]}) = ([X^*]\delta_{[\Bbbk]}, [Z]) = 0,$$

a contradiction. Conversely, if $\Bbbk \mid X \otimes X^*$, then $\Bbbk \mid X^{**} \otimes X^*$. This yields that

$$[X^*]\delta_{[\Bbbk]} = \delta_{[X^*]}.$$

We claim that $\Bbbk \mid Z \otimes Z^*$. Otherwise, $\delta_{[\Bbbk]}[Z] = 0$ by Theorem 1.4.6. Then

$$1 = (\delta_{[Z]}^*, [Z]) = (\delta_{[X^*]}, [Z]) = ([X^*], \delta_{[\Bbbk]}[Z]) = 0,$$

a contradiction.

(2) Applying Part (1) to the almost split sequence $0 \to Z^* \to Y^* \to X^* \to 0$, we may obtain the desired result. □

Denote by $\mathcal{J}_+$ and $\mathcal{J}_-$ the subgroups of $r(H)$ respectively as follows:

$$\mathcal{J}_+ := \mathbb{Z}\{\delta_{[M]} \mid [M] \in \text{ind}(H) \text{ and } \Bbbk \mid M \otimes M^*\}, \tag{1.13}$$

$$\mathcal{J}_- := \mathbb{Z}\{\delta_{[M]} \mid [M] \in \text{ind}(H) \text{ and } \Bbbk \mid M^* \otimes M\}. \tag{1.14}$$

By Theorem 1.4.6, $\mathcal{J}_+$ (resp. $\mathcal{J}_-$) is a right (resp. left) ideal of $r(H)$ generated by $\delta_{[\Bbbk]}$. Moreover, we have $\mathcal{J}_+^* = \mathcal{J}_-$ and $\mathcal{J}_-^* = \mathcal{J}_+$ by Proposition 1.4.8.

Now let $\mathcal{P}_+$ and $\mathcal{P}_-$ denote the subgroups of $r(H)$ as follows:

$$\mathcal{P}_+ := \mathbb{Z}\{[M] \in \text{ind}(H) \mid \Bbbk \nmid M \otimes M^*\}, \tag{1.15}$$

$$\mathcal{P}_- := \mathbb{Z}\{[M] \in \text{ind}(H) \mid \Bbbk \nmid M^* \otimes M\}. \tag{1.16}$$

Then $\mathcal{P}_+$ and $\mathcal{P}_-$ both contain the ideal $\mathcal{P}$ of $r(H)$. It follows from Proposition 1.2.8 that $\mathcal{P}_+$ is a right ideal of $r(H)$ and $\mathcal{P}_-$ is a left ideal of $r(H)$. Obviously, $\mathcal{P}_+^* = \mathcal{P}_-$ and $\mathcal{P}_-^* = \mathcal{P}_+$.

According to the associativity and non-degeneracy of the form $(-,-)$, we have $\mathcal{P}_- x = 0$ if and only if $(\mathcal{P}_-, x) = 0$ if and only if $(x, \mathcal{P}_-) = 0$ since $\mathcal{P}_- = \mathcal{P}_-^{**}$. Similarly, $x\mathcal{P}_+ = 0$ if and only if $(x, \mathcal{P}_+) = 0$ if and only if $(\mathcal{P}_+, x) = 0$. Thus, the right annihilator $r(\mathcal{P}_-)$ of $\mathcal{P}_-$ and the left annihilator $l(\mathcal{P}_+)$ of $\mathcal{P}_+$ can be expressed respectively as follows:

$$r(\mathcal{P}_-) := \{x \in r(H) \mid (x, y) = 0 \text{ for all } y \in \mathcal{P}_-\},$$

$$l(\mathcal{P}_+) := \{x \in r(H) \mid (y, x) = 0 \text{ for all } y \in \mathcal{P}_+\}.$$

The relationship between these one-sided ideals of $r(H)$ can be described as follows.

Proposition 1.4.9 *Let H be of finite representation type.*

(1) $\mathcal{J}_+ = r(\mathcal{P}_-)$.

(2) $\mathcal{J}_- = l(\mathcal{P}_+)$.

Proof It is sufficient to prove Part (1) and the proof of Part (2) is similar. For two indecomposable modules X and Y satisfying $\Bbbk \mid X \otimes X^*$ and $\Bbbk \nmid Y^* \otimes Y$, by Theorem 1.4.6, we have

$$(\delta_{[X]}, [Y]) = (\delta_{[\Bbbk]}[X], [Y]) = ([Y^{**}]\delta_{[\Bbbk]}, [X]) = (0, [X]) = 0.$$

This implies that $\mathcal{J}_+ \subseteq r(\mathcal{P}_-)$. For any $x \in r(\mathcal{P}_-)$,

$$\begin{aligned} x &= \sum_{[M]\in \mathrm{ind}(H)} (x, [M])\delta_{[M]}^* \quad \text{(by Remark 1.4.4 (1))} \\ &= \sum_{\Bbbk \mid M^*\otimes M} (x, [M])\delta_{[M]}^* \quad (\text{as } x \in r(\mathcal{P}_-)). \end{aligned}$$

We have that $x \in \mathcal{J}_-^* = \mathcal{J}_+$, and hence $r(\mathcal{P}_-) \subseteq \mathcal{J}_+$. □

Next, we shall use these one-sided ideals to get information about the Jacobson radical and central primitive idempotents of $r(H)$. We first need the following lemma:

Lemma 1.4.10 *For any $x \in r(H)$, we have*

(1) *If $xx^* = 0$, then $x \in \mathcal{P}_+$.*

(2) *If $x^*x = 0$, then $x \in \mathcal{P}_-$.*

Proof It suffices to prove Part (1), and the proof of Part (2) is similar. Suppose

$$x = \sum_{\Bbbk \mid M\otimes M^*} \lambda_{[M]}[M] + \sum_{\Bbbk \nmid M\otimes M^*} \lambda_{[M]}[M],$$

where each $\lambda_{[M]} \in \mathbb{Z}$. By Theorem 1.2.7 (1) and Corollary 1.4.7, the coefficient of the identity $[\Bbbk]$ in the linear expression of xx^* with respect to the basis $\mathrm{ind}(H)$ is $\sum_{\Bbbk|M\otimes M^*} \lambda_{[M]}^2$. Thus, if $xx^* = 0$, then $\lambda_{[M]} = 0$ for any indecomposable module M satisfying $\Bbbk \mid M \otimes M^*$. Hence

$$x = \sum_{\Bbbk\nmid M\otimes M^*} \lambda_{[M]}[M] \in \mathcal{P}_+. \qquad \square$$

Proposition 1.4.11 *Let H be of finite representation type. If the Green ring $r(H)$ is commutative, then the Jacobson radical $J(r(H))$ of $r(H)$ is contained in $\mathcal{P}_+ \cap \mathcal{P}_-$.*

Proof Since $r(H)$ is commutative and finitely generated as an algebra over $\mathbb{Z}$, the Jacobson radical $J(r(H))$ is equal to the nilradical of $r(H)$. For any $x \in J(r(H))$, let $x_0 := x$ and $x_{i+1} := x_i x_i^*$ for $i \geqslant 0$. Then there exists some k such that $x_k = 0$. Denote

$$x = \sum_{\Bbbk|M\otimes M^*} \lambda_{[M]}[M] + \sum_{\Bbbk\nmid M\otimes M^*} \lambda_{[M]}[M]$$

and

$$x_1 = xx^* = \sum_{\Bbbk|M\otimes M^*} \mu_{[M]}[M] + \sum_{\Bbbk\nmid M\otimes M^*} \mu_{[M]}[M],$$

for all $\lambda_{[M]}$ and $\mu_{[M]}$ in $\mathbb{Z}$. As shown in the proof of Lemma 1.4.10, the coefficient of $[\Bbbk]$ in $x_1 = xx^*$ is

$$\mu_{[\Bbbk]} = \sum_{\Bbbk|M\otimes M^*} \lambda_{[M]}^2$$

and the coefficient of $[\Bbbk]$ in $x_2 = x_1 x_1^*$ is

$$\sum_{\Bbbk|M\otimes M^*} \mu_{[M]}^2.$$

If $\mu_{[\Bbbk]} \neq 0$, then $\sum_{\Bbbk|M\otimes M^*} \mu_{[M]}^2 \neq 0$, and hence $x_2 \neq 0$. Repeating this process, we obtain that $x_i \neq 0$ for any $i \geqslant 0$. This contradicts to the fact that $x_k = 0$. In view of this, $\mu_{[\Bbbk]} = 0$, and hence

$$x = \sum_{\Bbbk\nmid M\otimes M^*} \lambda_{[M]}[M] \in \mathcal{P}_+.$$

Similarly, if $x \in J(r(H))$, then $x \in \mathcal{P}_-$. We obtain that

$$J(r(H)) \subseteq \mathcal{P}_+ \cap \mathcal{P}_-. \qquad \square$$

Now we are able to locate central primitive idempotents of $r(H)$.

Proposition 1.4.12 *Let e be a central primitive idempotent of $r(H)$. Then, either $e \in \mathcal{P}_+ \cap \mathcal{P}_-$ or $1 - e \in \mathcal{P}_+ \cap \mathcal{P}_-$.*

Proof If e is a central primitive idempotent of $r(H)$, so is e^* since the duality operator $*$ is an anti-automorphism of $r(H)$. It follows that $e = e^*$ or $ee^* = e^*e = 0$. If $ee^* = e^*e = 0$, by Lemma 1.4.10, $e \in \mathcal{P}_+$ and $e \in \mathcal{P}_-$ as well. Suppose $e = e^*$ and

$$e = \sum_{\Bbbk \mid M \otimes M^*} \lambda_{[M]}[M] + \sum_{\Bbbk \nmid M \otimes M^*} \lambda_{[M]}[M].$$

Comparing the coefficients of $[\Bbbk]$ in both sides of the equation $ee^* = e$, we obtain that

$$\sum_{\Bbbk \mid M \otimes M^*} \lambda_{[M]}^2 = \lambda_{[\Bbbk]}.$$

This implies that $\lambda_{[\Bbbk]} = 0$ or 1 and $\lambda_{[M]} = 0$ for all $[M]$ satisfying $[M] \neq [\Bbbk]$ and $\Bbbk \mid M \otimes M^*$. Hence e has the following reduced form:

$$e = \lambda_{[\Bbbk]}[\Bbbk] + \sum_{\Bbbk \nmid M \otimes M^*} \lambda_{[M]}[M].$$

In the meanwhile, if we write

$$e = \sum_{\Bbbk \mid M^* \otimes M} \mu_{[M]}[M] + \sum_{\Bbbk \nmid M^* \otimes M} \mu_{[M]}[M].$$

Then the equation $e^*e = e$ yields that

$$e = \mu_{[\Bbbk]}[\Bbbk] + \sum_{\Bbbk \nmid M^* \otimes M} \mu_{[M]}[M].$$

Thus, $\mu_{[\Bbbk]} = \lambda_{[\Bbbk]}$ which is equal to 0 or 1. We conclude that

$$e \in \mathcal{P}_+ \cap \mathcal{P}_- \text{ if } \mu_{[\Bbbk]} = \lambda_{[\Bbbk]} = 0, \quad 1 - e \in \mathcal{P}_+ \cap \mathcal{P}_- \text{ if } \mu_{[\Bbbk]} = \lambda_{[\Bbbk]} = 1.$$ □

Chapter 2 The Green Rings of Spherical Hopf Algebras

In this chapter, we devote ourselves to the study of the Green ring $r(H)$ of a spherical Hopf algebra H. We endow $r(H)$ with a new bilinear form. This form is shown to be associative and symmetric and its radical is the annihilator of a special central element of $r(H)$. After that we consider two quotients of $r(H)$, one of them is the Benson-Carlson quotient ring $r(H)/\mathcal{P}$, where $\mathcal{P}$ is the free abelian group generated by all indecomposable H-modules of quantum dimension zero. This quotient ring can be realized as the Green ring of a factor category of H-mod. If, moreover, H is of finite representation type, the complexified quotient ring admits group-like algebra as well as bi-Frobenius algebra structure.

2.1 A new bilinear form

In this section, we focus our attention on the Green ring of a finite dimensional spherical Hopf algebra. We will see that this Green ring can be endowed with an associative and symmetric bilinear form $\langle -, - \rangle$. We describe a relationship between the bilinear form $\langle -, - \rangle$ and the bilinear from $(-, -)$ described in (1.11). We show that the radical of the form $\langle -, - \rangle$ is the annihilator of the central element $1 - \delta^*_{[\Bbbk]}$.

Let H be a finite dimensional non-semisimple *spherical Hopf algebra* over a field $\Bbbk$. That is, there is a group-like element ω of H such that

$$S^2(h) = \omega h \omega^{-1}, \quad \text{for } h \in H, \tag{2.1}$$

$$\mathrm{tr}_X(\vartheta\omega) = \mathrm{tr}_X(\vartheta\omega^{-1}), \tag{2.2}$$

for any finite dimensional H-module X and any $\vartheta \in \mathrm{End}_H(X)$ (see [2, Section 2.1]). It follows from [2, Proposition 2.1] that the condition (2.2) is only required for any simple H-module. Applying $\vartheta = id_X$ to (2.2), one obtains the *quantum dimension* of X:

$$\mathbf{d}(X) := \mathrm{tr}_X(\omega) = \mathrm{tr}_X(\omega^{-1}).$$

Observe that

$$\mathbf{d}(X) = \mathbf{d}(X^*) \quad \text{and} \quad \mathbf{d}(X \otimes Y) = \mathbf{d}(X)\mathbf{d}(Y).$$

The map $\mathbf{d}$ from $r(H)$ to $\Bbbk$ given by quantum dimensions of H-modules is a ring homomorphism preserving the duality operator $*$.

Remark 2.1.1 *Note that the map $\theta : X \to X^{**}$ given by $\theta(x)(f) = f(\omega x)$ for $x \in X$ and $f \in X^*$ is an H-module isomorphism. The left quantum trace of the isomorphism θ defined in* (1.1) *is computed as follows:*

$$\mathrm{Tr}_X^L(\theta) = \mathrm{tr}_X(\omega) = \mathbf{d}(X).$$

Proposition 2.1.2 *Let X be an indecomposable H-module. The following statements are equivalent:*

(1) $\mathbf{d}(X) \neq 0$.

(2) $\Bbbk \mid X \otimes X^*$.

(3) $\Bbbk \mid X^* \otimes X$.

(4) $\delta_{[\Bbbk]}[X] = \delta_{[X]}$.

(5) $[X]\delta_{[\Bbbk]} = \delta_{[X]}$.

Proof (1) $\Leftrightarrow$ (2). Since $\mathbf{d}(X) = \mathrm{Tr}_X^L(\theta)$, it can be seen from Theorem 1.2.7 (see also [95, Theorem 2.4]) that the first two statements are equivalent.

(2) $\Leftrightarrow$ (4). It follows from Theorem 1.4.6 (4).

(3) $\Leftrightarrow$ (5). It follows from Theorem 1.4.6 (3).

(1) $\Leftrightarrow$ (3). Since Part (1) and Part (2) are equivalent, it follows that $\mathbf{d}(X^*) \neq 0$ if and only if $\Bbbk \mid X^* \otimes X^{**}$, if and only if $\Bbbk \mid X^* \otimes X$ since $X^{**} \cong X$. Note that $\mathbf{d}(X^*) = \mathbf{d}(X)$. Thus, $\mathbf{d}(X) \neq 0$ if and only if $\Bbbk \mid X^* \otimes X$. □

Together with Proposition 2.1.2 and Theorem 1.4.6, one obtains the following result:

Proposition 2.1.3 *Let X be an indecomposable H-module. The following statements are equivalent:*

(1) $\mathbf{d}(X) = 0$.

(2) $\Bbbk \nmid X \otimes X^*$.

(3) $\Bbbk \nmid X^* \otimes X$.

(4) $\delta_{[\Bbbk]}[X] = 0$.

(5) $[X]\delta_{[\Bbbk]} = 0$.

Remark 2.1.4 (1) *Proposition* 2.1.2 *together with* (1.13) *and* (1.14) *shows that $\mathcal{J}_+ = \mathcal{J}_-$, which is denoted by $\mathcal{J}$, so $\mathcal{J}$ is a two-sided ideal of $r(H)$ generated by the central element $\delta_{[\Bbbk]}$.*

(2) *Proposition* 2.1.3 *together with* (1.15) *and* (1.16) *shows that $\mathcal{P}_+ = \mathcal{P}_-$, which is denoted by $\mathcal{P}$, so $\mathcal{P}$ is a two-sided ideal of $r(H)$ generated by all H-modules of quantum dimension zero.*

(3) *Let H be of finite representation type. It follows from Proposition* 1.4.11 *that the Jacobson radical of $r(H)$ is contained in $\mathcal{P}$ if $r(H)$ is commutative. It is easy to*

see that $\mathcal{P}^{\perp}$, the orthogonal to $\mathcal{P}$ with respect to the bilinear form $(-,-)$ defined in (1.11), *is nothing but the ideal $\mathcal{J}$.*

Note that the category H-mod of finite dimensional left H-modules is a spherical category (we refer to [6] for this definition). For H-modules X and Y, there is a bilinear pairing given by

$$\Theta : \mathrm{Hom}_H(X,Y) \times \mathrm{Hom}_H(Y,X) \to \Bbbk, \quad \Theta(f,g) = \mathrm{tr}_X(\omega g f).$$

A morphism f from X to Y is called a *negligible morphism* if $\Theta(f,g) = 0$ for any morphism g from Y to X. The negligible morphisms form a monoidal ideal, i.e., composing or tensoring a negligible morphism with any morphism yields a negligible morphism [58, P.118]. This leads to a factor category H-$\underline{\text{mod}}$ whose objects are those of H-mod and the morphism spaces are the quotient:

$$\underline{\mathrm{Hom}}_H(X,Y) := \mathrm{Hom}_H(X,Y)/\mathcal{J}(X,Y),$$

where $\mathcal{J}(X,Y)$ is the set consisting of all negligible morphisms from X to Y. The factor category H-$\underline{\text{mod}}$ is an additive semisimple $\Bbbk$-linear spherical category [6] with the monoidal structure derived from that of H-mod.

We give a new bilinear form $\langle -,- \rangle$ on $r(H)$ defined by

$$\langle [X],[Y] \rangle := \dim_{\Bbbk} \mathcal{J}(X,Y^*).$$

To see the associativity and symmetry of the form, we need the following lemma.

Lemma 2.1.5 *For H-modules X, Y and Z, we have the following $\Bbbk$-linear isomorphisms*:

(1) $\mathcal{J}(X \otimes Y, Z) \cong \mathcal{J}(X, Z \otimes Y^*)$.

(2) $\mathcal{J}(X, Y \otimes Z) \cong \mathcal{J}(Y^* \otimes X, Z)$.

Proof We only prove Part (1) and the proof of Part (2) is similar. Consider the following canonical isomorphism (see [5, Lemma 2.1.6])

$$\Phi_{X,Y,Z} : \mathrm{Hom}_H(X \otimes Y, Z) \to \mathrm{Hom}_H(X, Z \otimes Y^*)$$

given by

$$\Phi_{X,Y,Z}(\alpha) = (\alpha \otimes id_{Y^*})(id_X \otimes \mathrm{coev}_Y),$$

where $\mathrm{coev}_Y : \Bbbk \to Y \otimes Y^*$ is the coevaluation of Y defined by $\mathrm{coev}_Y(1) = \sum_i y_i \otimes y_i^*$ for a basis $\{y_i\}$ of Y and its dual basis $\{y_i^*\}$ of Y^*.

Claim 1 $\Phi_{X,Y,Z}(\mathcal{J}(X \otimes Y, Z)) \subseteq \mathcal{J}(X, Z \otimes Y^*)$.

We denote

$$\theta_Y : Y \to Y^{**}, \quad \theta_Y(x)(f) = f(\omega x)$$

the pivotal structure of the spherical category H-mod. If $f \in \mathcal{J}(X \otimes Y, Z)$, for any $g \in \mathrm{Hom}_H(Z \otimes Y^*, X)$, the morphism $(id_X \otimes \theta_Y^{-1})\Phi_{Z,Y^*,X}(g)$ is in $\mathrm{Hom}_H(Z, X \otimes Y)$ and hence

$$\mathrm{tr}_Z(f(id_X \otimes \theta_Y^{-1})\Phi_{Z,Y^*,X}(g)\omega) = 0.$$

Next, we use this equality to prove that $\Phi_{X,Y,Z}(f) \in \mathcal{J}(X, Z \otimes Y^*)$, namely,

$$\mathrm{tr}_{Z\otimes Y^*}(\Phi_{X,Y,Z}(f)g\omega) = 0.$$

For any H-module M, let $\{m_i\}$ be a basis of M and $\{m_i^*\}$ the dual basis of M^*. We have

$$\sum_i \omega m_i \otimes \omega m_i^* = \sum_i m_i \otimes m_i^*,$$

or equivalently,

$$\sum_i m_i \otimes \omega m_i^* = \sum_i \omega^{-1} m_i \otimes m_i^*.$$

Now it is straightforward to check that the image of a basis $\{z_i\}$ of Z under the morphism $\omega f(id_X \otimes \theta_Y^{-1})\Phi_{Z,Y^*,X}(g)$ is

$$(\omega f(id_X \otimes \theta_Y^{-1})\Phi_{Z,Y^*,X}(g))(z_i) = \sum_j f(\omega g(z_i \otimes y_j^*) \otimes y_j).$$

It follows that

$$\begin{aligned} 0 &= \mathrm{tr}_Z(f(id_X \otimes \theta_Y^{-1})\Phi_{Z,Y^*,X}(g)\omega) \\ &= \mathrm{tr}_Z(\omega f(id_X \otimes \theta_Y^{-1})\Phi_{Z,Y^*,X}(g)) \\ &= \sum_{i,j} \langle z_i^*, f(\omega g(z_i \otimes y_j^*) \otimes y_j)\rangle. \end{aligned} \tag{2.3}$$

Note that the image of the basis $\{z_i \otimes y_k^*\}$ of $Z \otimes Y^*$ under the morphism $\omega\Phi_{X,Y,Z}(f)g$ is

$$(\omega\Phi_{X,Y,Z}(f)g)(z_i \otimes y_k^*) = \sum_j f(\omega g(z_i \otimes y_k^*) \otimes y_j) \otimes y_j^*.$$

Therefore,

$$\begin{aligned} \mathrm{tr}_{Z\otimes Y^*}(\Phi_{X,Y,Z}(f)g\omega) &= \mathrm{tr}_{Z\otimes Y^*}(\omega\Phi_{X,Y,Z}(f)g) \\ &= \sum_{i,k,j} \langle z_i^* \otimes y_k^{**}, f(\omega g(z_i \otimes y_k^*) \otimes y_j) \otimes y_j^*\rangle \\ &= \sum_{i,j} \langle z_i^*, f(\omega g(z_i \otimes y_j^*) \otimes y_j)\rangle \end{aligned}$$

$$= 0 \quad \text{(by (2.3))}.$$

Claim 2 $\Phi_{X,Y,Z}(\mathcal{J}(X \otimes Y, Z)) = \mathcal{J}(X, Z \otimes Y^*)$.

For any morphism

$$\alpha \in \mathcal{J}(X, Z \otimes Y^*) \subseteq \mathrm{Hom}_H(X, Z \otimes Y^*) = \Phi_{X,Y,Z}(\mathrm{Hom}_H(X \otimes Y, Z)),$$

there is some $f \in \mathrm{Hom}_H(X \otimes Y, Z)$ such that $\alpha = \Phi_{X,Y,Z}(f)$. Now we check that $f \in \mathcal{J}(X \otimes Y, Z)$, namely, for any $g \in \mathrm{Hom}_H(Z, X \otimes Y)$, the trace $\mathrm{tr}_Z(fg\omega)$ needs to be zero. Since the morphism $\Phi^{-1}_{Z,Y^*,X}((id_X \otimes \theta_Y)g)$ is in $\mathrm{Hom}_H(Z \otimes Y^*, X)$, we have

$$\mathrm{tr}_{Z\otimes Y^*}(\alpha\Phi^{-1}_{Z,Y^*,X}((id_X \otimes \theta_Y)g)\omega) = 0.$$

We write

$$g(z_i) = \sum_j x_{ij} \otimes y_j \in X \otimes Y,$$

then the image of the basis $\{z_i \otimes y_k^*\}$ of $Z \otimes Y^*$ under the morphism $\omega\alpha\Phi^{-1}_{Z,Y^*,X}((id_X \otimes \theta_Y)g)$ is

$$(\omega\alpha\Phi^{-1}_{Z,Y^*,X}((id_X \otimes \theta_Y)g))(z_i \otimes y_k^*) = \sum_{j,s} \langle y_k^*, \omega y_j\rangle \omega f(x_{ij} \otimes y_s) \otimes \omega y_s^*.$$

It follows that

$$\begin{aligned}
0 &= \mathrm{tr}_{Z\otimes Y^*}(\alpha\Phi^{-1}_{Z,Y^*,X}((id_X \otimes \theta_Y)g)\omega) \\
&= \mathrm{tr}_{Z\otimes Y^*}(\omega\alpha\Phi^{-1}_{Z,Y^*,X}((id_X \otimes \theta_Y)g)) \\
&= \sum_{i,k,j,s} \langle y_k^*, \omega y_j\rangle \langle z_i^* \otimes y_k^{**}, \omega f(x_{ij} \otimes y_s) \otimes \omega y_s^*)\rangle \\
&= \sum_{i,k,j,s} \langle y_k^*, \omega y_j\rangle \langle z_i^* \otimes y_k^{**}, \omega f(x_{ij} \otimes \omega^{-1} y_s) \otimes y_s^*)\rangle \\
&= \sum_{i,k,j} \langle y_k^*, \omega y_j\rangle \langle z_i^*, \omega f(x_{ij} \otimes \omega^{-1} y_k)\rangle \\
&= \sum_{i,k,j} \langle \omega y_k^*, \omega y_j\rangle \langle z_i^*, \omega f(x_{ij} \otimes y_k)\rangle \\
&= \sum_{i,k,j} \langle y_k^*, y_j\rangle \langle z_i^*, \omega f(x_{ij} \otimes y_k)\rangle \\
&= \sum_{i,j} \langle z_i^*, \omega f(x_{ij} \otimes y_j)\rangle. \qquad (2.4)
\end{aligned}$$

Now the image of the basis $\{z_i\}$ of Z under the morphism ωfg is given by

$$(\omega fg)(z_i) = \sum_j \omega f(x_{ij} \otimes y_j),$$

and by (2.4) we have

$$\operatorname{tr}_Z(fg\omega) = \operatorname{tr}_Z(\omega fg) = \sum_{i,j} \langle z_i^*, \omega f(x_{ij} \otimes y_j)\rangle = 0.$$

We complete the proof. □

Proposition 2.1.6 *The form $\langle -, - \rangle$ on $r(H)$ has the following properties:*

(1) *The form $\langle -, - \rangle$ is associative and symmetric.*

(2) *The forms $(-, -)$ and $\langle -, - \rangle$ satisfy the following relation:*

$$\langle [X], [Y] \rangle = ([X](1 - \delta^*_{[\Bbbk]}), [Y]).$$

Proof (1) The associativity of the form $\langle -, - \rangle$ follows from Lemma 2.1.5, i.e.,

$$\begin{aligned}\langle [X][Y], [Z] \rangle &= \dim_{\Bbbk} \mathcal{J}(X \otimes Y, Z^*)\\ &= \dim_{\Bbbk} \mathcal{J}(X, (Y \otimes Z)^*)\\ &= \langle [X], [Y][Z] \rangle.\end{aligned}$$

The $\Bbbk$-linear isomorphisms

$$\mathcal{J}(X, Y^*) \cong \mathcal{J}(Y^{**}, X^*) \cong \mathcal{J}(Y, X^*)$$

(the former isomorphism follows from Lemma 2.1.5) imply that

$$\langle [X], [Y] \rangle = \langle [Y], [X] \rangle.$$

(2) Note that the factor category H-$\underline{\text{mod}}$ is an additive semisimple $\Bbbk$-linear spherical category and there is one to one correspondence between the set of simple objects in H-$\underline{\text{mod}}$ and the set of indecomposable objects with non-zero quantum dimension in H-mod (see [2, Theorem 2.7]). For indecomposable H-modules X and Y, we have

$$([X], [Y]) - \langle [X], [Y] \rangle = \dim_{\Bbbk} \underline{\operatorname{Hom}}_H(X, Y^*) = \begin{cases} 1, & X \cong Y^* \text{ and } \mathbf{d}(X) \neq 0, \\ 0, & \text{otherwise.} \end{cases}$$

This shows that

$$\langle [X], [Y] \rangle = ([X], [Y]) - ([X]\delta^*_{[\Bbbk]}, [Y]) = ([X](1 - \delta^*_{[\Bbbk]}), [Y]),$$

as desired. □

The radical of the form $\langle -,-\rangle$ is the set consisting of $x \in r(H)$ such that $\langle x, y\rangle = 0$ for all $y \in r(H)$. It follows from Proposition 2.1.6 that x belongs to the radical of the form $\langle -,-\rangle$ if and only if $x(1-\delta^*_{[\Bbbk]}) = 0$ since the form $(-,-)$ is non-degenerate. Obviously, the form $\langle -,-\rangle$ is non-degenerate if the central element $1-\delta^*_{[\Bbbk]}$ is invertible.

Proposition 2.1.7 *Let* $0 \to \tau(\Bbbk) \to E \to \Bbbk \to 0$ *be the almost split sequence ending at the trivial module* $\Bbbk$*. The central element* $1-\delta^*_{[\Bbbk]}$ *is invertible with the inverse* $1-\delta_{[\Bbbk]}$ *if and only if* $[E] \in \mathcal{P}$.

Proof Applying the duality functor $*$ on the almost split sequence ending at the trivial module $\Bbbk$, we obtain the almost split sequence $0 \to \Bbbk \to E^* \to \tau(\Bbbk)^* \to 0$. This implies that $\delta^*_{[\Bbbk]} = \delta_{[\tau(\Bbbk)^*]}$. However, $\tau(\Bbbk)^*$ is indecomposable and $\mathbf{d}(\tau(\Bbbk)^*) \neq 0$, see Proposition 1.4.8. It follows from Proposition 2.1.2 that

$$\delta_{[\tau(\Bbbk)^*]} = \delta_{[\Bbbk]}[\tau(\Bbbk)^*].$$

Now $\delta_{[\Bbbk]}[\tau(\Bbbk)^*] = \delta^*_{[\Bbbk]}$ shows that

$$\begin{aligned}(1-\delta_{[\Bbbk]})(1-\delta^*_{[\Bbbk]}) &= (1-\delta_{[\Bbbk]})([E]^* - [\tau(\Bbbk)]^*)\\ &= [E]^* - [\tau(\Bbbk)]^* - \delta_{[\Bbbk]}[E]^* + \delta_{[\Bbbk]}[\tau(\Bbbk)]^*\\ &= ([E]^* - [\tau(\Bbbk)]^* + \delta^*_{[\Bbbk]}) - \delta_{[\Bbbk]}[E]^*\\ &= 1 - \delta_{[\Bbbk]}[E]^*.\end{aligned}$$

It follows from Proposition 2.1.2 and Proposition 2.1.3 that $\delta_{[\Bbbk]}[E]^* = 0$ if and only if $[E]^*$ is a linear expression of some indecomposable H-modules of quantum dimension zero, if and only if $[E] \in \mathcal{P}$. □

Example 2.1.8 For the Sweedler Hopf algebra $H = \Bbbk\langle x, g \mid x^2, g^2-1, gx+xg\rangle$, the Hopf algebra H is spherical and the middle term of the almost split sequence ending at the trivial module $\Bbbk$ belongs to $\mathcal{P}$ since it is projective [2, Example 2.5]. Hence $1-\delta^*_{[\Bbbk]}$ is invertible and the form $\langle -,-\rangle$ on $r(H)$ is non-degenerate.

2.2 Quotients of Green rings

In this section, we study the quotient rings $r(H)/\mathcal{P}$ and $r(H)/\mathcal{J}$ respectively. In particular, we show that the quotient ring $r(H)/\mathcal{P}$ can be regarded as the Green ring of the factor category H-$\underline{\text{mod}}$.

Denote by

$$\mathbf{B}_0 = \{[X] \in \text{ind}(H) \mid \mathbf{d}(X) = 0\},$$

$$\mathbf{B}_1 = \{[X] \in \text{ind}(H) \mid \mathbf{d}(X) \neq 0\}$$

and

$$\mathbf{\Delta} = \{\delta^*_{[X]} \mid [X] \in \mathbf{B}_0\}.$$

Then $\text{ind}(H) = \mathbf{B}_0 \cup \mathbf{B}_1$. Let $\mathbb{Z}\mathbf{B}_1$ (resp. $\mathbb{Z}\mathbf{\Delta}$) be the subgroup of $r(H)$ generated by the set $\mathbf{B}_1$ (resp. $\mathbf{\Delta}$).

Note that $\mathcal{P} = \mathbb{Z}\mathbf{B}_0$. The projection of $\text{ind}(H) = \mathbf{B}_0 \cup \mathbf{B}_1$ onto $\mathbf{B}_1$ induces an isomorphism $\mathbb{Z}\mathbf{B}_1 \cong r(H)/\mathcal{P}$ as abelian groups. This isomorphism transfers the obvious multiplication of $r(H)/\mathcal{P}$ to $\mathbb{Z}\mathbf{B}_1$ as follows:

$$x \cdot y = \sum_{[X] \in \mathbf{B}_1} (\delta^*_{[X]}, xy)[X], \quad \text{for } x, y \in \mathbb{Z}\mathbf{B}_1.$$

The associativity of the multiplication is clear. It can also be checked directly as follows:

$$\begin{aligned}
(x \cdot y) \cdot z &= \sum_{[X] \in \mathbf{B}_1} (\delta^*_{[X]}, xy)[X] \cdot z \\
&= \sum_{[X] \in \mathbf{B}_1} (\delta^*_{[X]}, xy) \sum_{[Y] \in \mathbf{B}_1} (\delta^*_{[Y]}, [X]z)[Y] \\
&= \sum_{[Y] \in \mathbf{B}_1} (\delta^*_{[Y]}, \sum_{[X] \in \mathbf{B}_1} (\delta^*_{[X]}, xy)[X]z)[Y] \\
&= \sum_{[Y] \in \mathbf{B}_1} (\delta^*_{[Y]}, xyz - \sum_{[X] \in \mathbf{B}_0} (\delta^*_{[X]}, xy)[X]z)[Y] \\
&= \sum_{[Y] \in \mathbf{B}_1} (\delta^*_{[Y]}, xyz)[Y].
\end{aligned} \tag{2.5}$$

The last two equalities in (2.5) hold because of the fact that $(\mathcal{J}, \mathcal{P}) = 0$ and the following identity, which follows from Lemma 1.4.2 (2):

$$xy = \sum_{[X] \in \mathbf{B}_0} (\delta^*_{[X]}, xy)[X] + \sum_{[X] \in \mathbf{B}_1} (\delta^*_{[X]}, xy)[X].$$

It is similar that

$$x \cdot (y \cdot z) = \sum_{[Y] \in \mathbf{B}_1} (\delta^*_{[Y]}, xyz)[Y].$$

The identity of $\mathbb{Z}\mathbf{B}_1$ is 1 since

$$x = \sum_{[X] \in \mathbf{B}_1} (\delta^*_{[X]}, x)[X], \quad \text{for any } x \in \mathbb{Z}\mathbf{B}_1.$$

Remark 2.2.1 *Repeating the process of* (2.5), *one obtains that*

$$x_1 \cdot x_2 \cdot \cdots \cdot x_m = \sum_{[X]\in \mathbf{B}_1} (\delta^*_{[X]}, x_1 x_2 \cdots x_m)[X].$$

Proposition 2.2.2 *The ring* $\mathbb{Z}\mathbf{B}_1$ *has the following properties:*
(1) *For* $x, y \in \mathbb{Z}\mathbf{B}_1$, *we have* $(x \cdot y)^* = y^* \cdot x^*$.
(2) *The map* $\mathbf{d} : \mathbb{Z}\mathbf{B}_1 \to \Bbbk$ *given by quantum dimension is a ring homomorphism.*

Proof (1) Note that $(\delta^*_{[\Bbbk]}, x)$ stands for the coefficient of $[\Bbbk]$ in the linear expression of x with respect to the basis $\mathbf{B}_1$. It follows that

$$(\delta^*_{[\Bbbk]}, x) = (\delta^*_{[\Bbbk]}, x^*).$$

This gives rise to the following result:

$$\begin{aligned}(x \cdot y)^* &= \sum_{[X]\in\mathbf{B}_1} (\delta^*_{[X]}, xy)[X]^* = \sum_{[X]\in\mathbf{B}_1} (\delta^*_{[\Bbbk]}, [X]^* xy)[X]^* \\ &= \sum_{[X]\in\mathbf{B}_1} (\delta^*_{[\Bbbk]}, [X]xy)[X] = \sum_{[X]\in\mathbf{B}_1} (\delta^*_{[\Bbbk]}, y^* x^* [X]^*)[X] \\ &= \sum_{[X]\in\mathbf{B}_1} (\delta^*_{[X]}, y^* x^*)[X] = y^* \cdot x^*.\end{aligned}$$

(2) Note that $\mathbf{d}([X]) = 0$ for $[X] \in \mathbf{B}_0$. For $x, y \in \mathbb{Z}\mathbf{B}_1$, we have

$$\begin{aligned}\mathbf{d}(x \cdot y) &= \sum_{[X]\in\mathbf{B}_1} (\delta^*_{[X]}, xy)\mathbf{d}([X]) \\ &= \sum_{[X]\in\mathbf{B}_1} (\delta^*_{[X]}, xy)\mathbf{d}([X]) + \sum_{[X]\in\mathbf{B}_0} (\delta^*_{[X]}, xy)\mathbf{d}([X]) \\ &= \mathbf{d}\Big(\sum_{[X]\in\mathbf{B}_1} (\delta^*_{[X]}, xy)[X] + \sum_{[X]\in\mathbf{B}_0} (\delta^*_{[X]}, xy)[X]\Big) \\ &= \mathbf{d}(xy) = \mathbf{d}(x)\mathbf{d}(y).\end{aligned}$$

The proof is completed. □

Next, we give an interpretation of $\mathbb{Z}\mathbf{B}_1$ from a categorical point of view. We show that $\mathbb{Z}\mathbf{B}_1$ is in fact the Green ring of the factor category H-$\underline{\text{mod}}$.

Theorem 2.2.3 *The Green ring of the factor category* H-$\underline{\text{mod}}$ *is isomorphic to* $\mathbb{Z}\mathbf{B}_1$.

Proof The canonical functor F from H-mod to H-$\underline{\text{mod}}$ given by $F(M) = M$ for any H-module M, and $F(\alpha) = \overline{\alpha}$ for $\alpha \in \mathrm{Hom}_H(M, N)$ with the canonical image $\overline{\alpha} \in \underline{\mathrm{Hom}}_H(M, N)$ is a full and dense tensor functor. Such a functor induces a ring epimorphism f from $r(H)$ to the Green ring of H-$\underline{\text{mod}}$ such that $f(\mathcal{P}) = 0$. Hence there is a unique ring epimorphism $\overline{f}$ from $r(H)/\mathcal{P}$ to the Green ring of H-$\underline{\text{mod}}$ such that $\overline{f}(\overline{x}) = f(x)$, for any $x \in r(H)$ with the canonical image $\overline{x} \in r(H)/\mathcal{P}$. The rank of the Green ring of H-$\underline{\text{mod}}$ is the same as that of $r(H)/\mathcal{P}$ since there is one to one correspondence between the set of simple objects in H-$\underline{\text{mod}}$ and the set of indecomposable objects with non-zero quantum dimension in H-mod [2, Theorem 2.7]. We conclude that the Green ring of H-$\underline{\text{mod}}$ is isomorphic to $r(H)/\mathcal{P}$ (or equivalently, $\mathbb{Z}\mathbf{B}_1$). □

Example 2.2.4 Suppose that char($\Bbbk$) = 0. Let G be a finite group, g a central element of G such that $g^2 \neq 1$ and χ a $\Bbbk$-linear character of G of order 2. Let H be the algebra generated by y and all h in G such that $\Bbbk G$ is a subalgebra of H and

$$y^2 = g^2 - 1, \ yh = \chi(h)hy, \quad \text{for } h \in G.$$

Then H is a spherical Hopf algebra with the comultiplication Δ, counit ε, and antipode S given respectively by

$$\Delta(y) = y \otimes g + 1 \otimes y, \quad \varepsilon(y) = 0, \quad S(y) = -yg^{-1},$$

$$\Delta(h) = h \otimes h, \quad \varepsilon(h) = 1, \quad S(h) = h^{-1},$$

for $h \in G$. It follows from [87] that any finite dimensional indecomposable H-module V is either simple or projective. If V is not projective, then

$$\mathbf{d}(V) = \chi_V(g) \dim_{\Bbbk}(V),$$

where χ_V is the character of V as a $\Bbbk G$-module. Let N be the normal subgroup of G generated by g^2 and $\overline{G}$ the quotient of G by N. The factor category H-$\underline{\text{mod}}$ is now equivalent to $\Bbbk\overline{G}$-mod and the Green ring $\mathbb{Z}\mathbf{B}_1$ of H-$\underline{\text{mod}}$ is isomorphic to the Grothendieck ring $\mathrm{G}_0(\Bbbk\overline{G})$ of $\Bbbk\overline{G}$.

Remark 2.2.5 *If H is of finite representation type, it is similar that the isomorphism $\mathbb{Z}\mathbf{\Delta} \cong r(H)/\mathcal{J}$ as abelian groups transfers the multiplication of $r(H)/\mathcal{J}$ to $\mathbb{Z}\mathbf{\Delta}$ as follows:*

$$x \star y = \sum_{[X] \in \mathbf{B}_0} (xy, [X])\delta^*_{[X]}, \quad \text{for } x, y \in \mathbb{Z}\mathbf{\Delta}. \tag{2.6}$$

The identity of $\mathbb{Z}\mathbf{\Delta}$ is $\sum_{[X]\in\mathbf{B}_0}(1, [X])\delta^*_{[X]}$. One can repeat the process of (2.6) and obtain that

$$x_1 \star x_2 \star \cdots \star x_m = \sum_{[X] \in \mathbf{B}_0} (x_1x_2 \cdots x_m, [X])\delta^*_{[X]}.$$

2.3 Group-like algebra and bi-Frobenius algebra structure

In this section, we show that when H is a finite dimensional spherical Hopf algebra of finite representation type, the algebra $\Bbbk\mathbf{B}_1$ is a group-like algebra as well as a bi-Frobenius algebra.

The notation of a group-like algebra was introduced by Doi in [21] generalizing the group algebra of a finite group and a scheme ring (Bose-Mesner algebra) of a non-commutative association scheme.

Definition 2.3.1 (cf. [21]) *Let $(A, \varepsilon, \mathbf{b}, *)$ be a quadruple, where A is a finite dimensional algebra over a field $\Bbbk$ with the unity 1, ε is an algebra morphism from A to $\Bbbk$, the set $\mathbf{b} = \{b_i \mid i \in I\}$ is a $\Bbbk$-basis of A such that $0 \in I$ and $b_0 = 1$, and $*$ is an involution of the index set I. Then $(A, \varepsilon, \mathbf{b}, *)$ is called a group-like algebra if the following conditions hold*:

(G1) $\varepsilon(b_i) = \varepsilon(b_{i^*}) \neq 0$ *for all* $i \in I$.

(G2) $p_{ij}^k = p_{j^*i^*}^{k^*}$ *for all* $i, j, k \in I$, *where* p_{ij}^k *is the structure constant for* $\mathbf{b}$ *defined by* $b_ib_j = \sum_{k\in I} p_{ij}^k b_k$.

(G3) $p_{ij}^0 = \delta_{i,j^*}\varepsilon(b_i)$ *for all* $i, j \in I$.

Group-like algebras have some special properties, see e.g., [21]. Group-like algebras of dimension 2 and 3 have been determined in [21]. For the study of group-like algebras of dimension 4, we refer to [22].

Definition 2.3.2 (cf. [22,23]) *A Frobenius coalgebra is a pair (C, t) where C is a finite dimensional coalgebra and $t \in C$ such that the morphism $t \leftharpoonup : C^* \to C,\ f \mapsto t \leftharpoonup f$ is bijective; or equivalently, the morphism $\rightharpoonup t : C^* \to C,\ f \mapsto f \rightharpoonup t$ is bijective, where $t \leftharpoonup f = f(t_{(1)})t_{(2)}$ and $f \rightharpoonup t = t_{(1)}f(t_{(2)})$.*

The notation of a Frobenius coalgebra has a nice characterization that is analogue to the characterization of a Frobenius algebras [20,21]. The concept of a bi-Frobenius algebra was introduced by Doi and Takeuchi in [23] and further investigated in [20,21] as a natural generalized of finite dimensional Hopf algebras.

Definition 2.3.3 (cf. [20]) *Let H be a finite dimensional algebra and coalgebra over a field $\Bbbk$, $\phi \in H^*$, $t \in H$. Define a map S by*

$$S : H \to H,\ S(x) = t \leftharpoonup (x \rightharpoonup \phi) = \phi(t_{(1)}x)t_{(2)}.$$

The quadruple (H, ϕ, t, S) is called a bi-Frobenius algebra if the following conditions hold:

(BF1) *The counit ε of the coalgebra H is an algebra morphism.*

(BF2) *The unity 1 is a group-like element of H.*

(BF3) *(H, ϕ) is a Frobenius algebra.*

(BF4) *(H, t) is a Frobenius coalgebra.*

(BF5) *S is an anti-algebra and anti-coalgebra morphism, i.e., $S(ab) = S(b)S(a)$, $S(1) = 1$ and $\Delta(S(a)) = S(a_{(2)}) \otimes S(a_{(1)})$, $\varepsilon(S(a)) = \varepsilon(a)$ for $a, b \in H$.*

The map S given above is necessarily bijective [23], it is called the *antipode* of the bi-Frobenius algebra H. The antipode S does not mean a convolution inverse of identity. This is true in the particular situation of Hopf algebras.

Example 2.3.4 Let H be a finite dimensional Hopf algebra. Choose a right integral $\lambda \in H^*$ and a left integral $\Lambda \in H$ such that $\lambda(\Lambda) = 1$. Then (H, λ, Λ, S) becomes a bi-Frobenius algebra.

Remark 2.3.5 (1) *Let* $(A, \varepsilon, \mathbf{b}, *)$ *be a group-like algebra. Then* A *becomes a coalgebra by defining* $\Delta(b_i) = \dfrac{1}{\varepsilon(b_i)} b_i \otimes b_i$. *Let* $\phi \in A^*$ *such that* $\phi(b_i) = \delta_{0,i}$ *and* $t = \sum_{i \in I} b_i$. *Define the* $\Bbbk$*-linear map* S *from* A *to itself given by* $S(b_i) = b_{i^*}$ *for any* $i \in I$. *Then* (A, ϕ, t, S) *becomes a bi-Frobenius algebra with a pair of dual bases* $\left\{ b_i, \dfrac{b_{i^*}}{\varepsilon(b_i)} \mid i \in I \right\}$ *(see* [21, *Remark* 3.2]*).*

(2) *If a group-like algebra is also a Hopf algebra, then it needs to be a group algebra* [37, *Corollary* 2]. *Because of this, a bi-Frobenius algebra coming from a group-like algebra is not a Hopf algebra if the algebra itself is not a group algebra.*

It is interesting to construct bi-Frobenius algebras that are not Hopf algebras. Using known results on the existence of large Hadamard matrices, the author in [37] constructed a class of bi-Frobenius algebras of arbitrarily large dimension satisfying the additional condition

$$S * id = id * S = u \circ \varepsilon, \tag{2.7}$$

and that are not Hopf algebras. This family of bi-Frobenius algebras satisfying the condition (2.7) is also studied in [71]. There are many other approaches to construct bi-Frobenius algebras that are not Hopf algebras, see e.g., [83, 92]. As we shall see, one of main results of this section is that $\Bbbk\mathbf{B}_1$ is a bi-Frobenius algebra that is not a Hopf algebra.

Consider the $\mathbb{Z}$-linear map from $\mathbb{Z}\mathbf{B}_1$ to $\mathbb{Z}$ given by $T(x) = (\delta^*_{[\Bbbk]}, x)$ for $x \in \mathbb{Z}\mathbf{B}_1$. The map T is in fact a trace function as shown in the following lemma:

Lemma 2.3.6 *For* $x, y \in \mathbb{Z}\mathbf{B}_1$, *the map* T *satisfies the following properties*:

(1) $T(x) = T(x^*)$.

(2) $T(xy) = T(yx)$ *and* $T(x \cdot y) = T(xy)$.

(3) *The* $\mathbb{Z}$*-bilinear form* $[-,-]$ *on* $\mathbb{Z}\mathbf{B}_1$ *given by* $[x, y] = T(xy)$ *is an associative, symmetric, non-degenerate and* $*$*-invariant bilinear form.*

Proof (1) It follows from the fact that $(\delta^*_{[\Bbbk]}, x) = (\delta^*_{[\Bbbk]}, x^*)$ for all $x \in \mathbb{Z}\mathbf{B}_1$.

(2) We have $T(xy) = T(yx)$ since the form $(-,-)$ is symmetric and $\delta^*_{[\Bbbk]}$ is a central element of $r(H)$. Note that

$$xy - x \cdot y = \sum_{[X] \in \mathbf{B}_0} (\delta^*_{[X]}, xy)[X].$$

Then $T(x \cdot y) = T(xy)$ since $T([X]) = 0$ for any $[X] \in \mathbf{B}_0$.

(3) The associativity of the form $[-,-]$ is obvious. The symmetry stems from $T(xy) = T(yx)$ and the non-degeneracy of the form follows from the orthogonality:

$$T([X][Y]^*) = \delta_{[X],[Y]}, \quad \text{for } [X], [Y] \in \mathbf{B}_1.$$

Moreover, this form is $*$-invariant:

$$[x^*, y^*] = T(x^*y^*) = T(yx) = T(xy) = [x, y],$$

as desired. □

Let $\Bbbk\mathbf{B}_1 := \Bbbk \otimes_{\mathbb{Z}} \mathbb{Z}\mathbf{B}_1$. The map $\mathbf{d}$ arises naturally an algebra morphism from $\Bbbk\mathbf{B}_1$ to $\Bbbk$. The left integral space of $\Bbbk\mathbf{B}_1$ with respect to $\mathbf{d}$ is the following set:

$$\int_{\Bbbk\mathbf{B}_1}^{l} := \{t \in \Bbbk\mathbf{B}_1 \mid x \cdot t = \mathbf{d}(x)t \text{ for } x \in \Bbbk\mathbf{B}_1\}.$$

Similarly, one can define the right integral space of $\Bbbk\mathbf{B}_1$ with respect to $\mathbf{d}$:

$$\int_{\Bbbk\mathbf{B}_1}^{r} := \{t \in \Bbbk\mathbf{B}_1 \mid t \cdot x = \mathbf{d}(x)t \text{ for } x \in \Bbbk\mathbf{B}_1\}.$$

If the spaces of left integral and right integral coincide, then $\Bbbk\mathbf{B}_1$ is called *unimodular*.

In the sequel, we assume that H is of finite representation type. In this case, the algebra $\Bbbk\mathbf{B}_1$ is a finite dimensional symmetric algebra with a pair of dual bases $\{[X], [X]^* \mid [X] \in \mathbf{B}_1\}$ with respect to the bilinear form $[-,-]$. Moreover, it is unimodular with $\int_{\Bbbk\mathbf{B}_1}^{l}$ as well as $\int_{\Bbbk\mathbf{B}_1}^{r}$ spanned by

$$t = \sum_{[X] \in \mathbf{B}_1} \mathbf{d}([X])[X]^*.$$

We denote

$$\mathbf{b} = \{x := \mathbf{d}([X])[X] \mid [X] \in \mathbf{B}_1\}.$$

Then $\mathbf{b}$ is a basis of $\Bbbk\mathbf{B}_1$.

Proposition 2.3.7 *The quadruple* $(\Bbbk\mathbf{B}_1, \mathbf{d}, \mathbf{b}, *)$ *is a group-like algebra.*

Proof We verify the conditions (G1)—(G3) given in Definition 2.3.1. The condition (G1) is obvious. To verify the condition (G2), we suppose that

$$x \cdot y = \sum_{z \in \mathbf{b}} p_{xy}^{z} z, \tag{2.8}$$

where $p_{xy}^z \in \Bbbk$. On the one hand, applying the duality operator $*$ to the equality (2.8), we obtain that

$$y^* \cdot x^* = \sum_{z \in \mathbf{b}} p_{xy}^z z^*.$$

On the other hand, we have

$$y^* \cdot x^* = \sum_{z \in \mathbf{b}} p_{y^*x^*}^z z.$$

Thus, $p_{xy}^z = p_{y^*x^*}^{z^*}$ for $x, y, z \in \mathbf{b}$. Now we verify the condition (G3). Extending the map $T : \mathbb{Z}\mathbf{B}_1 \to \mathbb{Z}$ to $T : \Bbbk\mathbf{B} \to \Bbbk$ by linearity, one has for $x, y \in \mathbf{b}$ that

$$\begin{aligned}
p_{xy}^1 &= T(x \cdot y) \\
&= \mathbf{d}([X])\mathbf{d}([Y])T([X] \cdot [Y]) \\
&= \mathbf{d}([X])\mathbf{d}([Y])T([X][Y]) \\
&= \mathbf{d}([X])\mathbf{d}([Y])(\delta_{[\Bbbk]}^*, [X][Y]) \\
&= \mathbf{d}([X])\mathbf{d}([Y])(\delta_{[X^*]}^*, [Y]) \\
&= \mathbf{d}([X])\mathbf{d}([Y])\delta_{[X^*],[Y]} \\
&= \delta_{[X^*],[Y]}\mathbf{d}(x).
\end{aligned}$$

Therefore, the condition (G3) is satisfied. □

A group-like algebra can be viewed as a bi-Frobenius algebra in a natural way, see Example 2.3.5. Following this approach, we define on $(\Bbbk\mathbf{B}_1, \mathbf{d}, \mathbf{b}, *)$ a bi-Frobenius algebra structure as follows:

$(\Bbbk\mathbf{B}_1, \phi)$ is a Frobenius algebra with the Frobenius homomorphism ϕ given by $\phi(x) = \delta_{x,1}$ for $x \in \mathbf{b}$. Equivalently,

$$\phi([X]) = \begin{cases} 1, & [X] = 1, \\ 0, & [X] \neq 1. \end{cases}$$

The set $\{x, \dfrac{x^*}{\mathbf{d}(x)} \mid x \in \mathbf{b}\}$ forms a pair of dual bases of $(\Bbbk\mathbf{B}_1, \phi)$. This is equivalent to saying that $\{[X], [X]^*] \mid [X] \in \mathbf{B}_1\}$ is a pair of dual bases of $\Bbbk\mathbf{B}_1$ with respect to the Frobenius homomorphism ϕ. From the observation above, we conclude that the Frobenius homomorphism ϕ of $\Bbbk\mathbf{B}_1$ is nothing but the map T.

$\Bbbk\mathbf{B}_1$ is a coalgebra with the counit given by $\mathbf{d}$, and the comultiplication Δ defined by

$$\Delta(x) = \frac{1}{\mathbf{d}(x)} x \otimes x, \quad \text{for } x \in \mathbf{b}.$$

Equivalently,

$$\Delta([X]) = \frac{1}{\mathbf{d}([X])}[X] \otimes [X],$$

for $[X] \in \mathbf{B}_1$. Let $t = \sum_{[X]\in\mathbf{B}_1} \mathbf{d}([X])[X]$ (t is exactly an integral of $\Bbbk\mathbf{B}_1$ associated to the counit $\mathbf{d}$). Then $(\Bbbk\mathbf{B}_1, t)$ is a Frobenius coalgebra. Define the map $S : \Bbbk\mathbf{B}_1 \to \Bbbk\mathbf{B}_1$ by $S(x) = x^*$. It is easy to see that the map S is an anti-algebra and anti-coalgebra morphism, so is an antipode of $\Bbbk\mathbf{B}_1$. Now the quadruple $(\Bbbk\mathbf{B}_1, \phi, t, S)$ forms a bi-Frobenius algebra. We refer to [20–23] for more results on group-like algebras and bi-Frobenius algebras.

Chapter 3 The Stable Green Rings of Hopf Algebras

Throughout this chapter, H is an arbitrary finite dimensional Hopf algebra over an algebraically closed field $\Bbbk$. The category of finite dimensional left H-modules is denoted by H-mod. In this chapter, we study the Green ring of the stable category of H-mod. We provide a new bilinear form on the Green ring of the stable category. If this bilinear form is non-degenerate and H is a finite dimensional non-semisimple Hopf algebra of finite representation type, we show that the complexified Green algebra is a group-like algebra, and hence a bi-Frobenius algebra.

3.1 Stable Green rings

In this section, we shall give a definition of the stable Green ring of a finite dimensional Hopf algebra H, we show that the previous bilinear form $(-,-)$ on the Green ring $r(H)$ could induce a bilinear form on the stable Green ring of H. The induced form on the stable Green ring is associative but degenerate in general. We give some equivalent conditions for the non-degeneracy of the form.

Recall that the *stable category* H-$\underline{\text{mod}}$ has the same objects as H-mod does, and the space of morphisms from X to Y in H-$\underline{\text{mod}}$ is the following quotient space:

$$\underline{\text{Hom}}_H(X,Y) := \text{Hom}_H(X,Y)/\mathcal{P}(X,Y),$$

where $\mathcal{P}(X,Y)$ is the subspace of $\text{Hom}_H(X,Y)$ consisting of morphisms factoring through projective modules. The stable category H-$\underline{\text{mod}}$ is a triangulated [38] monoidal category with the monoidal structure stemming from that of H-mod.

Proposition 3.1.1 *The stable category H-$\underline{\text{mod}}$ is semisimple if and only if any indecomposable H-module is either simple or projective.*

Proof If any indecomposable H-module is either simple or projective, using the same method as [2, Theorem 2.7], one is able to prove that H-$\underline{\text{mod}}$ is semisimple. Conversely, suppose that H-$\underline{\text{mod}}$ is semisimple. Note that all simple objects of H-$\underline{\text{mod}}$ are those non-projective indecomposable H-modules. If H-mod has an indecomposable object M which is neither simple nor projective, then the indecomposable H-modules M and $\text{Soc}M$ are two simple objects in H-$\underline{\text{mod}}$. Since the inclusion map $\text{Soc}M \to M$ induces a surjective map $M^* \to (\text{Soc}M)^*$, it follows from Proposition 1.3.10 (2) that

$$\dim_{\Bbbk} \underline{\mathrm{Hom}}_H(M^*, (\mathrm{Soc}M)^*) = \langle M^*, (\mathrm{Soc}M)^*\rangle_3 = \langle M^*, (\mathrm{Soc}M)^*\rangle_1 \neq 0.$$

This means that $M^* \cong (\mathrm{Soc}M)^*$ in H-$\underline{\mathrm{mod}}$, so is an isomorphism in H-mod [77, Ch III, Lemma 4.3], a contradiction. □

The Green ring of the stable category H-$\underline{\mathrm{mod}}$ is called the *stable Green ring* of H, denoted by $r_{st}(H)$. Obviously, the stable Green ring $r_{st}(H)$ admits a $\mathbb{Z}$-basis consisting of all isomorphism classes of indecomposable non-projective H-modules. As the stable category H-$\underline{\mathrm{mod}}$ is a quotient category of H-mod, the stale Green ring $r_{st}(H)$ can be regarded as the quotient ring of the Green ring $r(H)$.

Proposition 3.1.2 *The stable Green ring $r_{st}(H)$ is isomorphic to the quotient ring $r(H)/\mathcal{P}$, where $\mathcal{P}$ is the ideal of $r(H)$ generated by all projective H-modules.*

Proof The canonical functor F from H-mod to H-$\underline{\mathrm{mod}}$ given by $F(M) = M$ and $F(\phi) = \underline{\phi}$, for $\phi \in \mathrm{Hom}_H(M, N)$ with the canonical image $\underline{\phi} \in \underline{\mathrm{Hom}}(M, N)$, is a full dense tensor functor. Such a functor induces a ring epimorphism f from $r(H)$ to $r_{st}(H)$ such that $f(\mathcal{P}) = 0$. Hence there is a unique ring epimorphism $\overline{f}$ from $r(H)/\mathcal{P}$ to $r_{st}(H)$ such that $\overline{f}(\overline{x}) = f(x)$, for any $x \in r(H)$ with the canonical image $\overline{x} \in r(H)/\mathcal{P}$. For two H-modules M and N without nonzero projective direct summands, it follows from [77, Ch III, Lemma 4.3] that $M \cong N$ in H-mod if and only if $M \cong N$ in H-$\underline{\mathrm{mod}}$. From this we conclude that $r_{st}(H)$ is isomorphic to $r(H)/\mathcal{P}$, since there is a one to one correspondence between the indecomposable objects in H-$\underline{\mathrm{mod}}$ and the non-projective indecomposable objects in H-mod. □

We identify $r(H)/\mathcal{P}$ with $r_{st}(H)$ and denote $\overline{x}$ the element in $r_{st}(H)$ for any $x \in r(H)$. Since $(\delta^*_{[\Bbbk]}, x) = 0$ for any $x \in \mathcal{P}$, the linear functional $(\delta^*_{[\Bbbk]}, -)$ on $r(H)$ induces a linear functional on $r_{st}(H)$. Using this functional, we define a new bilinear form on $r_{st}(H)$ as follows:

$$[\overline{x}, \overline{y}]_{st} := (\delta^*_{[\Bbbk]}, xy), \quad \text{for } x, y \in r(H).$$

It is obvious that the form $[-,-]_{st}$ is associative and $*$-symmetric:

$$[\overline{x}, \overline{y}]_{st} = [\overline{y}^*, \overline{x}^*]_{st}.$$

The left radical of the form $[-,-]_{st}$ is the subgroup of $r_{st}(H)$ consisting of $\overline{x} \in r_{st}(H)$ such that $[\overline{x}, \overline{y}]_{st} = 0$ for all $\overline{y} \in r_{st}(H)$. The right radical of the form $[-,-]_{st}$ is defined similarly. The form $[-,-]_{st}$ is called non-degenerate if the left radical (or equivalently, the right radical) of the form $[-,-]_{st}$ is zero.

Proposition 3.1.3 *The left radical of the form $[-,-]_{st}$ is equal to $\mathcal{P}_+/\mathcal{P}$ and the right radical of the form $[-,-]_{st}$ is equal to $\mathcal{P}_-/\mathcal{P}$, where $\mathcal{P}_+$ and $\mathcal{P}_-$ are defined in* (1.15) *and* (1.16) *respectively.*

Proof We only consider the left radical of the form $[-,-]_{st}$. For $x, y \in r(H)$, if $x \in \mathcal{P}_+$, then $xy \in \mathcal{P}_+$ since $\mathcal{P}_+$ is a right ideal of $r(H)$. It follows that

$$[\overline{x}, \overline{y}]_{st} = (\delta^*_{[\Bbbk]}, xy) = 0,$$

and hence $\overline{x}$ belongs to the left radical of the form $[-,-]_{st}$. Conversely, for any

$$x = \sum_{[M]\in \operatorname{ind}(H)} \lambda_{[M]}[M],$$

we suppose that $\overline{x}$ belongs to the left radical of the form $[-,-]_{st}$. The inverse of $*$ under the composition is denoted by $\star$. For any $[M] \in \operatorname{ind}(H)$, by Theorem 1.4.6, we have

$$\begin{aligned} 0 &= [\overline{x}, \overline{[M]^\star}]_{st} = (\delta^*_{[\Bbbk]}, x[M]^\star) \\ &= ([M]^{***}\delta^*_{[\Bbbk]}, x) = ((\delta_{[\Bbbk]}[M])^*, x) \\ &= \begin{cases} 0, & \Bbbk \nmid M \otimes M^*, \\ \lambda_{[M]}, & \Bbbk \mid M \otimes M^*. \end{cases} \end{aligned}$$

This implies that

$$x = \sum_{\Bbbk \nmid M\otimes M^*} \lambda_{[M]}[M] \in \mathcal{P}_+.$$

The proof is completed. □

Now let $\mathcal{J}$ be the subgroup of $r(H)$ defined as follows:

$$\mathcal{J} = \mathbb{Z}\{\delta_{[M]} \mid [M] \in \operatorname{ind}(H) \text{ and } M \text{ not projective}\}.$$

Then $\mathcal{J}_+$ and $\mathcal{J}_-$ defined in (1.13) and (1.14) respectively are both contained in $\mathcal{J}$. If H is of finite representation type, then $\mathcal{J}$ is nothing but $\ker\varphi$ $(= \mathcal{P}^\perp)$ by Lemma 1.3.2 (5). We are now ready to characterize the non-degeneracy of the form $[-,-]_{st}$ by Proposition 3.1.3.

Proposition 3.1.4 *The following statements are equivalent:*

(1) *The form* $[-,-]_{st}$ *is non-degenerate.*

(2) $\mathcal{P}_+ = \mathcal{P}_- = \mathcal{P}$.

(3) $\mathcal{J}_+ = \mathcal{J}_- = \mathcal{J}$.

(4) $\mathcal{J}$ *is an ideal of* $r(H)$ *generated by the central element* $\delta_{[\Bbbk]}$, *the left annihilator* $l(\mathcal{J})$ *and the right annihilator* $r(\mathcal{J})$ *of* $\mathcal{J}$ *are both equal to* $\mathcal{P}$.

Proof It can be seen from Proposition 3.1.3 that Part (1) and Part (2) are equivalent. The equality $\mathcal{P}_+ = \mathcal{P}$ is equivalent to saying that $\Bbbk \nmid M \otimes M^*$ if and only if M is projective, or equivalently, $\Bbbk \mid M\otimes M^*$ if and only if M is not projective, this is precisely $\mathcal{J}_+ = \mathcal{J}$. Similarly, $\mathcal{P}_- = \mathcal{P}$ if and only if $\mathcal{J}_- = \mathcal{J}$.

(1) $\Rightarrow$ (4). If the form $[-,-]_{st}$ is non-degenerate, then $\mathcal{J}_+ = \mathcal{J}_- = \mathcal{J}$. It follows from Theorem 1.4.6 that $\delta_{[\Bbbk]}$ is a central element of $r(H)$ and $\mathcal{J}$ is an ideal of $r(H)$ generated by $\delta_{[\Bbbk]}$. Observe that $\mathcal{J}_+ = \mathcal{J}_- = \mathcal{J}$ implying that $\mathcal{J}^* = \mathcal{J}_-^* = \mathcal{J}_+ = \mathcal{J}$. This deduces that the left and right annihilators of $\mathcal{J}$ coincide: $l(\mathcal{J}) = r(\mathcal{J})$. Let $I := l(\mathcal{J}) = r(\mathcal{J})$. We claim that $I = \mathcal{P}$. The inclusion $\mathcal{P} \subseteq I$ is obvious. We denote

$$T_{st} = \{\overline{x} \in r_{st}(H) \mid [\overline{x}, 1]_{st} = 0\} \quad \text{and} \quad T = \{x \in r(H) \mid \overline{x} \in T_{st}\}.$$

Then $I \subseteq T$ since $\mathcal{J}x = 0$ if and only if $(\mathcal{J}, x) = 0$. Now I is an ideal of $r(H)$ satisfying $\mathcal{P} \subseteq I \subseteq T$. So $I/\mathcal{P}$ is an ideal of $r_{st}(H)$ contained in $T/\mathcal{P} = T_{st}$. However, T_{st} contains no nonzero proper ideals of $r_{st}(H)$ since the form $[-,-]_{st}$ is non-degenerate. This implies that $I = \mathcal{P}$.

(4) $\Rightarrow$ (1). If $[\overline{y}, \overline{x}]_{st} = 0$ for any $y \in r(H)$, then $[\overline{x}^*, \overline{y}^*]_{st} = 0$ since the form is $*$-symmetric. We have

$$0 = [\overline{x}^*, \overline{y}^*]_{st} = (\delta_{[\Bbbk]}^*, x^*y^*) = ((x\delta_{[\Bbbk]})^*, y^*),$$

for any $y \in r(H)$. Thus, $x\delta_{[\Bbbk]} = 0$, so $x \in l(\mathcal{J}) = \mathcal{P}$, and hence $\overline{x} = 0$. Similarly, if $[\overline{x}, \overline{y}]_{st} = 0$ for any $y \in r(H)$, then $\overline{x} = 0$. □

Remark 3.1.5 *If the form $[-,-]_{st}$ is non-degenerate, then the equality $\mathcal{J}_+ = \mathcal{J}$ implies that $\Bbbk \mid M\otimes M^*$ for any indecomposable non-projective module M. It deduces that $M \cong M^{**}$ by Theorem 1.2.7 (1). In this case, the operator $*$ on $r_{st}(H)$ is an involution.*

Under certain assumptions we are able to obtain further information about the Jacobson radical of $r(H)$ described below.

Theorem 3.1.6 *Let H be of finite representation type such that the Green ring $r(H)$ is commutative and the form $[-,-]_{st}$ on $r_{st}(H)$ is non-degenerate. The Jacobson radical $J(r(H))$ of $r(H)$ is equal to $\mathcal{P} \cap \mathcal{P}^{\perp}$ if and only if $G_0(H)$ is semiprime.*

Proof If $J(r(H)) = \mathcal{P} \cap \mathcal{P}^{\perp}$, it is obvious that $G_0(H)$ is semiprime, since $r(H)/\mathcal{P}^{\perp} \cong G_0(H)$ and the Jacobson radical $J(r(H))$ is the nilradical of $r(H)$. Conversely, the non-degeneracy of the form $[-,-]_{st}$ on $r_{st}(H)$ shows that $\mathcal{P}_+ = \mathcal{P}_- = \mathcal{P}$. This implies that $J(r(H)) \subseteq \mathcal{P}$ by Proposition 1.4.11. If $G_0(H)$ is semiprime, then the isomorphism $G_0(H) \cong r(H)/\mathcal{P}^{\perp}$ implies that $J(r(H)) \subseteq \mathcal{P}^{\perp}$, so we obtain that $J(r(H)) \subseteq \mathcal{P} \cap \mathcal{P}^{\perp}$. The inclusion $\mathcal{P} \cap \mathcal{P}^{\perp} \subseteq J(r(H))$ is obvious, since any element of $\mathcal{P} \cap \mathcal{P}^{\perp}$ has square zero which can be deduced from the non-degeneracy of the form $(-,-)$. □

Remark 3.1.7 *The map $\varphi : r(H) \to G_0(H)$ given in (1.8) restricting to the ideal $\mathcal{P}$ gives rise to the Cartan map $\varphi|_{\mathcal{P}} : \mathcal{P} \to G_0(H)$, whose kernel is exactly* $\ker(\varphi|_{\mathcal{P}}) = \mathcal{P} \cap \ker\varphi = \mathcal{P} \cap \mathcal{P}^{\perp}$.

Example 3.1.8 If H is a finite dimensional pointed Hopf algebra of rank one (e.g., Taft algebras [15], generalized Taft algebras [49] and Radford Hopf algebras [87]), then $G_0(H)$ is semiprime and the form $[-,-]_{st}$ on $r_{st}(H)$ is non-degenerate since $\mathcal{P}_+ = \mathcal{P}_- = \mathcal{P}$. It follows that $J(r(H)) = \mathcal{P} \cap \mathcal{P}^\perp = \ker(\varphi|_\mathcal{P})$, which is a principal ideal generated by a special element, see [86, 87] for details.

3.2 Bi-Frobenius algebra structure

In this section, we assume that H is a finite dimensional non-semisimple Hopf algebra of finite representation type and the form $[-,-]_{st}$ on $r_{st}(H)$ is non-degenerate. In this case, we show that the complexified stable Green algebra $R_{st}(H) := \mathbb{C} \otimes_\mathbb{Z} r_{st}(H)$ admits a group-like algebra structure, hence it is a bi-Frobenius algebra.

Let $\{[X_i] \mid i \in \mathbb{I}\}$ be the set of all isomorphism classes of finite dimensional non-projective indecomposable H-modules. Since the trivial module $\mathbb{k}$ is not projective, we may set $0 \in \mathbb{I}$ such that $[X_0] := [\mathbb{k}]$. Note that X is not projective if and only if X^* is not projective. Thus, the duality functor $*$ of H-mod induces an involution (see Remark 3.1.5) on the index set $\mathbb{I}$ defined by $[X_{i^*}] := [X_i^*]$ for any $i \in \mathbb{I}$.

Proposition 3.2.1 *The stable Green ring $r_{st}(H)$ is a transitive fusion ring with respect to the basis $\{\overline{[X_i]} \mid i \in \mathbb{I}\}$.*

Proof It is straightforward to verify that $r_{st}(H)$ satisfies the conditions of a fusion ring defined in [28, Definition 3.1.7], where the group homomorphism τ from $r_{st}(H)$ to $\mathbb{Z}$ is determined by $\tau(\overline{x}) = (\delta^*_{[\mathbb{k}]}, x)$ for any $\overline{x} \in r_{st}(H)$. The stable Green ring $r_{st}(H)$ is transitive (see [28, Definition 3.3.1]), namely, for $i, j \in \mathbb{I}$, there exist $k, l \in \mathbb{I}$ such that $\overline{[X_j][X_k]}$ and $\overline{[X_l][X_j]}$ contain $\overline{[X_i]}$ with a nonzero coefficient. In fact, we have

$$\mathbb{k} \mid X_j \otimes X_j^* \quad \text{since} \quad \mathcal{P}_+ = \mathcal{P}_- = \mathcal{P}.$$

This implies that $X_i \mid X_j \otimes X_j^* \otimes X_i$. Then we may find an indecomposable non-projective module X_k in $X_j^* \otimes X_i$ such that $X_i \mid X_j \otimes X_k$. Similarly, $X_i \mid X_i \otimes X_j^* \otimes X_j$, then there exists some X_l in $X_i \otimes X_j^*$ such that $X_i \mid X_l \otimes X_j$. □

Remark 3.2.2 *The stable Green ring $r_{st}(H)$ is a fusion ring under the condition that the form $[-,-]_{st}$ on $r_{st}(H)$ is non-degenerate. However, the stable category H-$\underline{mod}$ is not necessary semisimple by Proposition* 3.1.1. *A typical example is that the stable category of the Taft algebra of dimension n^2 for $n > 2$ is not semisimple, while the stable Green ring of the Taft algebra is a fusion ring.*

The fact that $r_{st}(H)$ is a transitive fusion ring enables us to define the Frobenius-Perron dimension of $\overline{[X_i]}$ for any $i \in \mathbb{I}$. Let $\mathrm{FPdim}(\overline{[X_i]})$ be the maximal non-negative eigenvalue of the matrix of left multiplication by $\overline{[X_i]}$ with respect to the basis $\{\overline{[X_i]} \mid i \in \mathbb{I}\}$ of $r_{st}(H)$. Then $\mathrm{FPdim}(\overline{[X_i]})$ is called the *Frobenius-Perron dimension* of $\overline{[X_i]}$. Extending FPdim linearly from the basis $\{\overline{[X_i]} \mid i \in \mathbb{I}\}$ of $r_{st}(H)$

to $R_{st}(H)$, we obtain a functional $\mathrm{FPdim}: R_{st}(H) \to \mathbb{C}$. The functional FPdim has the following properties, see Proposition 3.3.4, Proposition 3.3.6 and Proposition 3.3.9 in [28].

Proposition 3.2.3 *For any $i \in \mathbb{I}$, we have*

(1) $\mathrm{FPdim}(\overline{[X_i]}) \geqslant 1$.

(2) *The functional* $\mathrm{FPdim}: R_{st}(H) \to \mathbb{C}$ *is an algebra homomorphism.*

(3) $\mathrm{FPdim}(\overline{[X_i]}) = \mathrm{FPdim}(\overline{[X_{i^*}]})$.

Let $x_i := \mathrm{FPdim}(\overline{[X_i]})\overline{[X_i]}$ for any $i \in \mathbb{I}$. Then $\mathbf{b} = \{x_i \mid i \in \mathbb{I}\}$ is a basis of $R_{st}(H)$.

Theorem 3.2.4 *The quadruple $(R_{st}(H), \mathrm{FPdim}, \mathbf{b}, *)$ is a group-like algebra.*

Proof We need to verify the conditions (G1)—(G3) given in Definition 2.3.1. The condition (G1) is obvious. To verify the condition (G2), we have

$$x_i^* = \mathrm{FPdim}(\overline{[X_i]})(\overline{[X_i]})^* = \mathrm{FPdim}(\overline{[X_{i^*}]})\overline{[X_{i^*}]} = x_{i^*}. \tag{3.1}$$

Now for $i, j \in \mathbb{I}$, we suppose that

$$x_i x_j = \sum_{k \in \mathbb{I}} p_{ij}^k x_k, \tag{3.2}$$

where $p_{ij}^k \in \mathbb{C}$. On the one hand, applying the duality operator $*$ to the equality (3.2) and using (3.1), we obtain that

$$x_{j^*} x_{i^*} = \sum_{k \in \mathbb{I}} p_{ij}^k x_{k^*}.$$

On the other hand, we have

$$x_{j^*} x_{i^*} = \sum_{l \in \mathbb{I}} p_{j^* i^*}^l x_l.$$

It follows that $p_{ij}^k = p_{j^* i^*}^{k^*}$ for $i, j, k \in \mathbb{I}$. Now we verify the condition (G3):

$$\begin{aligned}
p_{ij}^0 &= \mathrm{FPdim}(\overline{[X_i]})\mathrm{FPdim}(\overline{[X_j]})(\delta_{[\Bbbk]}^*, [X_i][X_j]) \\
&= \mathrm{FPdim}(\overline{[X_i]})\mathrm{FPdim}(\overline{[X_j]})([X_j]^{**}\delta_{[\Bbbk]}^*, [X_i]) \\
&= \mathrm{FPdim}(\overline{[X_i]})\mathrm{FPdim}(\overline{[X_j]})(\delta_{[X_{j^*}]}^*, [X_i]) \\
&= \mathrm{FPdim}(\overline{[X_i]})\mathrm{FPdim}(\overline{[X_j]})\delta_{i,j^*} \\
&= \delta_{i,j^*}\mathrm{FPdim}(x_i).
\end{aligned}$$

Therefore, the condition (G3) is satisfied. □

As noted in Remark 2.3.5, a group-like algebra is a bi-Frobenius algebra. Now let us look at the bi-Frobenius algebra structure induced from the group-like algebra structure on $R_{st}(H)$.

$(R_{st}(H), \phi)$ is a Frobenius algebra with the Frobenius homomorphism ϕ given by $\phi(x_i) = \delta_{0,i}$, for $i \in \mathbb{I}$. Equivalently,

$$\phi(\overline{[X_i]}) = \begin{cases} 1, & i = 0, \\ 0, & i \neq 0. \end{cases}$$

The set $\{x_i, \dfrac{x_i^*}{\mathrm{FPdim}(x_i)} \mid i \in \mathbb{I}\}$ forms a pair of dual bases of $(R_{st}(H), \phi)$. This is equivalent to saying that $\{\overline{[X_i]}, \overline{[X_{i^*}]} \mid i \in \mathbb{I}\}$ is a pair of dual bases of $(R_{st}(H), \phi)$. From the observation above, we conclude that the Frobenius homomorphism ϕ is nothing but the map determined by the form $[-,-]_{st}$, namely, $\phi(\overline{x}) = [\overline{x}, 1]_{st}$ for $\overline{x} \in R_{st}(H)$.

The stable Green algebra $R_{st}(H)$ is a coalgebra with the counit given by FPdim, and the comultiplication Δ defined by

$$\Delta(x_i) = \frac{1}{\mathrm{FPdim}(x_i)} x_i \otimes x_i,$$

or equivalently,

$$\Delta(\overline{[X_i]}) = \frac{1}{\mathrm{FPdim}(\overline{[X_i]})} \overline{[X_i]} \otimes \overline{[X_i]}, \quad \text{for } i \in \mathbb{I}.$$

Let

$$t = \sum_{i \in \mathbb{I}} x_i = \sum_{i \in \mathbb{I}} \mathrm{FPdim}(\overline{[X_i]}) \overline{[X_i]}.$$

Then t is an integral of $R_{st}(H)$ associated to the counit FPdim. Now $(R_{st}(H), t)$ becomes a Frobenius coalgebra. Define a map

$$S : R_{st}(H) \to R_{st}(H), \quad \text{by } S(x_i) = x_{i^*},$$

namely,

$$S(\overline{[X_i]}) = \overline{[X_{i^*}]}, \quad \text{for } i \in \mathbb{I}.$$

The map S is exactly the duality operator $*$ on $R_{st}(H)$. It is an anti-algebra and anti-coalgebra morphism, so is an antipode of $R_{st}(H)$. Now the quadruple $(R_{st}(H), \phi, t, S)$ forms a bi-Frobenius algebra which is in general not a Hopf algebra.

3.3 Applications to Radford Hopf algebras

In this section, we consider a special finite dimensional pointed Hopf algebra of rank one, known as a Radford Hopf algebra. We describe the bi-Frobenius algebra structure on the complexified stable Green algebra from the polynomial point of view.

Given two integers $m > 1$ and $n > 1$. Let ω be a primitive mn-th root of unity and H an algebra over a field $\Bbbk$ of characteristic zero generated by g and y subject to relations

$$g^{mn} = 1, \quad yg = \omega^{-m}gy, \quad y^n = g^n - 1.$$

Then H is a Hopf algebra whose comultiplication Δ, counit ε, and antipode S are given respectively by

$$\Delta(y) = y \otimes g + 1 \otimes y, \quad \varepsilon(y) = 0, \quad S(y) = -yg^{-1},$$

$$\Delta(g) = g \otimes g, \quad \varepsilon(g) = 1, \quad S(g) = g^{-1}.$$

The Hopf algebra H is called a Radford Hopf algebra, which was introduced by Radford [70] so as to give an example of Hopf algebra whose Jacobson radical is not a Hopf ideal.

The Green ring and the stable Green ring of the Radford Hopf algebra H can be presented by generators and relations. Let $\mathbb{Z}[Y, Z, X_1, X_2, \cdots, X_{m-1}]$ be a polynomial ring over $\mathbb{Z}$ in variables $Y, Z, X_1, X_2, \cdots, X_{m-1}$. The Green ring $r(H)$ of H is isomorphic to the quotient ring of $\mathbb{Z}[Y, Z, X_1, X_2, \cdots, X_{m-1}]$ modulo the ideal generated by the following elements from (3.3) to (3.5) (see [87, Theorem 8.2]):

$$Y^n - 1, \quad (1 + Y - Z)F_n(Y, Z), \quad YX_1 - X_1, \quad ZX_1 - 2X_1, \tag{3.3}$$

$$X_1^j - n^{j-1}X_j, \quad \text{for } 1 \leqslant j \leqslant m-1, \tag{3.4}$$

$$X_1^m - n^{m-2}(1 + Y + \cdots + Y^{n-1})F_n(Y, Z), \tag{3.5}$$

where $F_n(Y, Z)$ is a Dickson polynomial of the second kind defined recursively by

$$F_1(Y, Z) = 1, \quad F_2(Y, Z) = Z \quad \text{and} \quad F_k(Y, Z) = ZF_{k-1}(Y, Z) - YF_{k-2}(Y, Z),$$

for $k \geqslant 3$. More precisely, the polynomial $F_k(y, z)$ can be expressed as follows (see [15, Lemma 3.11]):

$$F_k(Y, Z) = \sum_{i=0}^{[\frac{k-1}{2}]} (-1)^i \binom{k-1-i}{i} Y^i Z^{k-1-2i}.$$

The Grothendieck ring $G_0(H)$ is isomorphic to the ring of $\mathbb{Z}[Y, X_1, \cdots, X_{m-1}]$ modulo the ideal generated by

$$Y^n - 1, \quad YX_1 - X_1, \quad X_1^j - n^{j-1}X_j, \quad \text{for } 1 \leqslant j \leqslant m-1,$$

$$X_1^m - n^{m-1}(1 + Y + \cdots + Y^{n-1}),$$

see [87, Corollary 8.3] for details.

The stable Green ring $r_{st}(H)$ of H is isomorphic to the stable Green ring of the Taft algebra of dimension n^2 (see [87, Section 7]), while the latter is isomorphic to the quotient ring $\mathbb{Z}[Y, Z]/I$, where I is an ideal of $\mathbb{Z}[Y, Z]$ generated by $Y^n - 1$ and $F_n(Y, Z)$ (see [86, Proposition 6.1]).

The form $[-, -]_{st}$ on $r_{st}(H)$ is non-degenerate (see Example 3.1.8). As shown in the previous section, there is a bi-Frobenius algebra structure on the complexified stable Green algebra $\mathbb{C} \otimes_{\mathbb{Z}} r_{st}(H) \cong \mathbb{C}[Y, Z]/I$. Next, we shall describe the bi-Frobenius algebra structure on $\mathbb{C}[Y, Z]/I$ using a new basis rather than the canonical basis consisting of indecomposable non-projective H-modules. We need the following inverse version of Dickson polynomials, which can be found in [86, Lemma 6.4]:

Lemma 3.3.1 *For any $j \geqslant 1$, we have*

$$Z^j = \sum_{k=0}^{[\frac{j}{2}]} \binom{j}{k} \frac{j+1-2k}{j+1-k} Y^k F_{j+1-2k}(Y, Z).$$

Denote by $y^i z^j$ the image of $Y^i Z^j$ under the canonical map $\mathbb{C}[Y, Z] \to \mathbb{C}[Y, Z]/I$. Then the set $\{y^i z^j \mid 0 \leqslant i \leqslant n-1, 0 \leqslant j \leqslant n-2\}$ forms a basis of $\mathbb{C}[Y, Z]/I$. By Lemma 3.3.1, the following equation holds in $\mathbb{C}[Y, Z]/I$:

$$y^i z^j = \sum_{k=0}^{[\frac{j}{2}]} \binom{j}{k} \frac{j+1-2k}{j+1-k} y^{i+k} F_{j+1-2k}(y, z).$$

Thus, $\{y^i F_j(y, z) \mid 0 \leqslant i \leqslant n-1, 1 \leqslant j \leqslant n-1\}$ is a basis of $\mathbb{C}[Y, Z]/I$. Next, we shall use this basis to describe the bi-Frobenius algebra structure on $\mathbb{C}[Y, Z]/I$. Following from [86, Remark 4.4 (3)] we have

$$y^i F_j(y, z) y^k F_l(y, z) = \sum_{t=\zeta(j,l)}^{\min\{j,l\}-1} y^{i+k+t} F_{j+l-1-2t}(y, z), \tag{3.6}$$

where we set $\zeta(j, l) = 0$ if $j + l - 1 < n$, and $\zeta(j, l) = j + l - n$ if $j + l - 1 \geqslant n$.

Define two maps $\varepsilon : \mathbb{C}[Y, Z]/I \to \mathbb{C}$ and $\Delta : \mathbb{C}[Y, Z]/I \to \mathbb{C}[Y, Z]/I \otimes \mathbb{C}[Y, Z]/I$ by

$$\varepsilon(y^i F_j(y, z)) = F_j(1, 2\cos\frac{\pi}{n})$$

and

$$\Delta(y^i F_j(y, z)) = \frac{1}{F_j(1, 2\cos\frac{\pi}{n})} y^i F_j(y, z) \otimes y^i F_j(y, z).$$

Then both ε and Δ are well-defined since $F_n\left(1, 2\cos\frac{\pi}{n}\right) = 0$ (see [87, Theorem 7.3]). Moreover, it is straightforward to check that

$$(\Delta \otimes id) \circ \Delta = (id \otimes \Delta) \circ \Delta$$

and

$$(id \otimes \varepsilon) \circ \Delta = id = (\varepsilon \otimes id) \circ \Delta.$$

Hence $(\mathbb{C}[Y, Z]/I, \Delta, \varepsilon)$ is a coalgebra.

Define the linear map $\phi : \mathbb{C}[Y, Z]/I \to \mathbb{C}$ by

$$\phi(y^i F_j(y, z)) = \begin{cases} 1, & i = 0, j = 1, \\ 0, & \text{otherwise.} \end{cases}$$

Then $(\mathbb{C}[Y, Z]/I, \phi)$ is a Frobenius algebra and

$$\{y^i F_j(y, z), y^{1-i-j} F_j(y, z) \mid 0 \leqslant i \leqslant n-1, 1 \leqslant j \leqslant n-1\}$$

is a pair of dual bases of $\mathbb{C}[Y, Z]/I$ with respect to the Frobenius homomorphism ϕ.

Denote

$$t = \sum_{i=0}^{n-1} \sum_{j=1}^{n-1} F_j(1, 2\cos\frac{\pi}{n}) y^i F_j(y, z).$$

Then

$$\Delta(t) = t_{(1)} \otimes t_{(2)} = \sum_{i=0}^{n-1} \sum_{j=1}^{n-1} y^i F_j(y, z) \otimes y^i F_j(y, z).$$

Define the linear map $S : \mathbb{C}[Y, Z]/I \to \mathbb{C}[Y, Z]/I$ by

$$S(f) = \phi(t_{(1)} f) t_{(2)} = \sum_{i=0}^{n-1} \sum_{j=1}^{n-1} \phi(y^i F_j(y, z) f) y^i F_j(y, z).$$

We have the following result:

Theorem 3.3.2 *The quadruple* $(\mathbb{C}[Y, Z]/I, \phi, t, S)$ *is a bi-Frobenius algebra.*

Proof To prove that $(\mathbb{C}[Y, Z]/I, \phi, t, S)$ is a bi-Frobenius algebra, we only need to show that $\Delta(1) = 1 \otimes 1$, the counit ε is an algebra morphism and the map S is an anti-algebra as well as anti-coalgebra automorphism by [20, Lemma 1.2]. Indeed, the former two conclusions are obviously true. By the definition of S and the equality (3.6), we have

$$S(y^k F_l(y, z)) = \sum_{i=0}^{n-1} \sum_{j=1}^{n-1} \phi(y^i F_j(y, z) y^k F_l(y, z)) y^i F_j(y, z)$$

$$= \sum_{i=0}^{n-1} \sum_{j=1}^{n-1} \phi \left(\sum_{t=\varsigma(j,l)}^{\min\{j,l\}-1} y^{i+k+t} F_{j+l-1-2t}(y,z) \right) y^i F_j(y,z).$$

By the definition of ϕ, we have

$$\phi(y^{i+k+t} F_{j+l-1-2t}(y,z)) = 1$$

if and only if i, j and t satisfy $n \mid i+k+t$ and $j+l-1-2t=1$. Note that $t \leqslant \min\{j,l\}-1$. The equality $j+l-1-2t=1$ implies that $j=l$. In this case, $t=l-1$ and $n \mid i+k+l-1$. It follows that

$$S(y^k F_l(y,z)) = y^{1-k-l} F_l(y,z),$$

and S maps the basis to its dual basis, hence the map S is bijective. In particular, $S(1)=1$ and

$$\begin{aligned} S(y^i F_j(y,z) y^k F_l(y,z)) &= S \left(\sum_{t=\varsigma(j,l)}^{\min\{j,l\}-1} y^{i+k+t} F_{j+l-1-2t}(y,z) \right) \\ &= \sum_{t=\varsigma(j,l)}^{\min\{j,l\}-1} y^{1-(i+k+t)-(j+l-1-2t)} F_{j+l-1-2t}(y,z) \\ &= y^{1-i-j} F_j(y,z) y^{1-k-l} F_l(y,z) \\ &= S(y^i F_j(y,z)) S(y^k F_l(y,z)). \end{aligned}$$

We conclude that S is an anti-algebra map since $\mathbb{C}[Y,Z]/I$ is a commutative algebra. In addition,

$$(\varepsilon \circ S)(y^i F_j(y,z)) = \varepsilon(y^{1-i-j} F_j(y,z)) = \varepsilon(y^i F_j(y,z))$$

and

$$\begin{aligned} (\Delta \circ S)(y^i F_j(y,z)) &= \Delta(y^{1-i-j} F_j(y,z)) \\ &= \frac{1}{F_j(1, 2\cos\frac{\pi}{n})} y^{1-i-j} F_j(y,z) \otimes y^{1-i-j} F_j(y,z) \\ &= ((S \otimes S)\Delta^{\mathrm{op}})(y^i F_j(y,z)). \end{aligned}$$

It follows that S is an anti-coalgebra map on $\mathbb{C}[Y,Z]/I$. □

Remark 3.3.3 *Note that $\{y^i z^j \mid 0 \leqslant i \leqslant n-1, 0 \leqslant j \leqslant n-2\}$ is a basis of $\mathbb{C}[Y,Z]/I$. Using this basis we are able to describe the bi-Frobenius algebra structure on $(\mathbb{C}[Y,Z]/I, \phi, t, S)$ as follows:*

- $\Delta(y^i z^j) = \sum_{k=0}^{\lfloor \frac{j}{2} \rfloor} \binom{j}{k} \frac{j+1-2k}{(j+1-k)F_{j+1-2k}(1, 2\cos\frac{\pi}{n})} y^{i+k} F_{j+1-2k}(y, z)$
$\otimes y^{i+k} F_{j+1-2k}(y, z)$;
- $\phi(y^i z^j) = \begin{cases} \dbinom{j}{\frac{j}{2}} \dfrac{2}{j+2}, & 2 \mid j \text{ and } n \mid i + \frac{j}{2}, \\ 0, & \text{otherwise}; \end{cases}$
- $t = \sum_{i=0}^{n-1} \sum_{j=1}^{n-1} F_j\left(1, 2\cos\frac{\pi}{n}\right) y^i F_j(y, z)$;
- $S(y^i z^j) = \sum_{k=0}^{\lfloor \frac{j}{2} \rfloor} \binom{j}{k} \frac{j+1-2k}{j+1-k} y^{k-i-j} F_{j+1-2k}(y, z)$.

Chapter 4 The Casimir Numbers of Green Rings

Throughout this chapter, H is a finite dimensional Hopf algebra which is of finite representation type over an algebraically closed field $\Bbbk$. This chapter deals with the question of when the Green ring $r(H)$, or the Green algebra $r(H) \otimes_{\mathbb{Z}} K$ over a field K, is semisimple (namely, has zero Jacobson radical). To solve this question, we endow $r(H)$ with the Casimir number and show that $r(H) \otimes_{\mathbb{Z}} K$ is semisimple if and only if the Casimir number of $r(H)$ is not zero in K. For the Green ring $r(H)$ itself, $r(H)$ is semisimple if and only if the Casimir number of $r(H)$ is not zero. Then we focus on the cases where $H = \Bbbk G$ for a cyclic group G of order p over a field $\Bbbk$ of characteristic p. In this case, the Casimir number of the Green ring $r(\Bbbk G)$ is $2p^2$. This leads to a complete description of the Jacobson radical of the Green algebra $r(\Bbbk G) \otimes_{\mathbb{Z}} K$ for any field K.

4.1 The Jacobson semisimplicity of Green rings

In this section, we will give the definition of the Casimir number of the Green ring $r(H)$ for any finite dimensional Hopf algebra H of finite representation type. Using the Casimir number one is able to determine when the Green ring $r(H)$, or the Green algebra $r(H) \otimes_{\mathbb{Z}} K$ over any field K is semisimple (with zero Jacobson radical).

Let H be a finite dimensional Hopf algebra of finite representation type over an algebraically closed field $\Bbbk$. Recall that for any indecomposable H-module Z, if Z is not projective, there exists a unique almost split sequence $0 \to X \to Y \to Z \to 0$ with the ending term Z, we denote by $\delta_{[Z]}$ the element $[X] - [Y] + [Z]$ in $r(H)$; if Z is projective, we write $\delta_{[Z]} = [Z] - [\mathrm{rad}Z]$, where $\mathrm{rad}Z$ is the radical of Z.

For any $[Z] \in \mathrm{ind}(H)$, where $\mathrm{ind}(H)$ is the set of all isomorphism classes of finite dimensional indecomposable H-modules, we denote by $\delta^*_{[Z]}$ the image of $\delta_{[Z]}$ under the duality operator $*$ of $r(H)$. Since H is of finite representation type, the Green ring $r(H)$ is a Frobenius $\mathbb{Z}$-algebra (see Proposition 1.4.3) with the bilinear form $(-,-)$ given in (1.11), and $\{\delta^*_{[Z]}, [Z] \mid [Z] \in \mathrm{ind}(H)\}$ forms a pair of dual bases of $r(H)$ with respect to the form $(-,-)$. The *Casimir operator* of $r(H)$ is the map c

from $r(H)$ to its center $Z(r(H))$ given by

$$c(x)=\sum_{[Z]\in\mathrm{ind}(H)}[Z]x\delta^*_{[Z]},\quad \text{for } x\in r(H).$$

Here, $\sum_{[Z]\in\mathrm{ind}(H)}[Z]x\delta^*_{[Z]}$ is a central element of $r(H)$, for this we refer to [52, Section 3.1]. The *Casimir element* of $r(H)$ is $c(1)=\sum_{[Z]\in\mathrm{ind}(H)}[Z]\delta^*_{[Z]}$. In particular,

$$\dim_{\Bbbk}(c(1))=\dim_{\Bbbk}(H).$$

The *Casimir number* of $r(H)$ is defined to be the non-negative integer m satisfying

$$\mathbb{Z}\cap\mathrm{Im}c=(m).$$

Remark 4.1.1 *Let H and H' be two finite dimensional Hopf algebras of finite representation type. If H-mod and H'-mod are monoidally equivalent under a monoidal functor, then this functor induces an isomorphism $r(H)\cong r(H')$ preserving structure coefficients with respect to the basis* $\mathrm{ind}(H)$ *of* $r(H)$ *and the basis* $\mathrm{ind}(H')$ *of* $r(H')$. *Meanwhile, the Casimir number does not depend on the choice of an associative and non-degenerate bilinear form on* $r(H)$. *This means that the Casimir number of* $r(H)$ *is a monoidal invariant of* H*-mod.*

The Casimir number of $r(H)$ can be used to determine whether or not the Green algebra $r(H)\otimes_{\mathbb{Z}}K$ over a field K is semisimple.

Theorem 4.1.2 *The Green algebra $r(H)\otimes_{\mathbb{Z}}K$ over a field K is semisimple if and only if the Casimir number of $r(H)$ is not zero in K.*

Proof If $K=\mathbb{F}_p$, the finite field consisting of p elements, then the algebra $r(H)\otimes_{\mathbb{Z}}\mathbb{F}_p$ is separable (i.e., semisimple) if and only if $(p)\not\supseteq\mathrm{Im}c\cap\mathbb{Z}$, see [52, Proposition 6]. If $K=\mathbb{Q}$, then $r(H)\otimes_{\mathbb{Z}}\mathbb{Q}$ is separable if and only if $\mathrm{Im}c=Z(r(H))$ by Higman's theorem [39, Theorem 1], or equivalently, $c(x)=1$ for some $x\in r(H)\otimes_{\mathbb{Z}}\mathbb{Q}$. This is equivalent to say that $c(mx)=m$, where m is a positive integer such that $mx\in r(H)$. Precisely, $\mathbb{Z}\cap\mathrm{Im}c\neq(0)$. For a general field K, since $\mathbb{Q}$ (resp. $\mathbb{F}_p$) is a perfect field, any field extension $\mathbb{Q}\subseteq K$ (resp. $\mathbb{F}_p\subseteq K$) is separable. This implies that $r(H)\otimes_{\mathbb{Z}}K$ is semisimple if and only if $r(H)\otimes_{\mathbb{Z}}\mathbb{Q}$ (resp. $r(H)\otimes_{\mathbb{Z}}\mathbb{F}_p$) is semisimple. The proof is completed. □

Let $\mathbb{Z}\cap\mathrm{Im}c=(m)$. There exists some $x\in r(H)$ such that $c(x)=m$. Applying dimension to this equality, we have

$$m=\dim_{\Bbbk}(c(x))=\dim_{\Bbbk}(x)\dim_{\Bbbk}(c(1))=\dim_{\Bbbk}(x)\dim_{\Bbbk}(H).$$

It means that the Casimir number m of $r(H)$ is divisible by $\dim_{\Bbbk}(H)$. This is the result of [52, Proposition 22(a)]. In particular, we have the following corollary:

Corollary 4.1.3 *For any prime number p, if $p\mid\dim_{\Bbbk}(H)$, then $r(H)\otimes_{\mathbb{Z}}K$ is not semisimple over a field K of characteristic p.*

Remark 4.1.4 *Let H be a Hopf algebra such that the Green ring $r(H)$ is isomorphic to a group ring $\mathbb{Z}G$ for a finite group G. In this case, $\mathbb{Z} \cap \mathrm{Imc} = (m)$, where m is exactly the order of G. It follows from Theorem 4.1.2 that $\mathbb{Z}G \otimes_{\mathbb{Z}} K = KG$ is semisimple if and only if m is not zero in K. This is exactly the well-known Maschke's theorem. From this point of view, Theorem 4.1.2 can be viewed as the Green ring version of the Maschke's theorem.*

An interesting result is that the Casimir number of $r(H)$ can also be used to determine when the Green ring $r(H)$ itself is semisimple. To see this, we need the following lemma:

Lemma 4.1.5 *Let $J(r(H))$ be the Jacobson radical of $r(H)$ and $pr(H)$ the ideal of $r(H)$ generated by a prime number p.*

(1) *We have $(J(r(H))^n \subseteq pr(H)$ for some integer n.*

(2) *If $\mathbb{Z} \cap \mathrm{Imc} = (m)$ and $p \nmid m$, then $J(r(H)) \subseteq pr(H)$.*

Proof (1) The ring isomorphism $r(H)/pr(H) \cong r(H) \otimes_{\mathbb{Z}} \mathbb{F}_p$ shows that the quotient $r(H)/pr(H)$ is a finite ring. So the Jacobson radical $J(r(H)/pr(H))$ of $r(H)/pr(H)$ is nilpotent (see [54, Proposition IV.7]). The canonical ring epimorphism

$$\pi : r(H) \to r(H)/pr(H)$$

yields that

$$\pi(J(r(H))) \subseteq J(r(H)/pr(H)),$$

so $\pi(J(r(H)))$ is nilpotent. Thus, there exists a positive integer n such that $(J(r(H)))^n$ is contained in the kernel of π, namely,

$$(J(r(H)))^n \subseteq pr(H).$$

(2) If the prime number p satisfies $p \nmid m$, then $r(H) \otimes_{\mathbb{Z}} \mathbb{F}_p$ is semisimple by Theorem 4.1.2. In this case,

$$\pi(J(r(H))) \subseteq J(r(H)/pr(H)) = 0.$$

This implies that $J(r(H)) \subseteq pr(H)$. □

Theorem 4.1.6 *The Green ring $r(H)$ is semisimple (with zero Jacobson radical) if and only if the Casimir number of $r(H)$ is not zero.*

Proof Assume that the Jacobson radical $J(r(H))$ of $r(H)$ is zero. Consider the finite dimensional algebra $r(H) \otimes_{\mathbb{Z}} \mathbb{Q}$ over $\mathbb{Q}$. We first show that the Jacobson radical $J(r(H) \otimes_{\mathbb{Z}} \mathbb{Q})$ of $r(H) \otimes_{\mathbb{Z}} \mathbb{Q}$ is zero. For any $x \in J(r(H) \otimes_{\mathbb{Z}} \mathbb{Q})$, there exists a nonzero integer n such that

$$nx \in r(H) \cap J(r(H) \otimes_{\mathbb{Z}} \mathbb{Q}).$$

For $y, z \in r(H)$, we have

$$y(nx)z \in r(H) \cap J(r(H) \otimes_{\mathbb{Z}} \mathbb{Q}).$$

Since $J(r(H) \otimes_{\mathbb{Z}} \mathbb{Q})$ is nilpotent, $1 - y(nx)z$ is invertible in $r(H)$. This means that $nx \in J(r(H)) = 0$, and hence $x = 0$. Now $J(r(H) \otimes_{\mathbb{Z}} \mathbb{Q}) = 0$ and the algebra $r(H) \otimes_{\mathbb{Z}} \mathbb{Q}$ is semisimple, it follows from Theorem 4.1.2 that the Casimir number of $r(H)$ is not zero in $\mathbb{Q}$, so it is a nonzero integer. Conversely, if the Casimir number of $r(H)$ is $m \neq 0$, then the set Ω consisting of all prime numbers p such that $p \nmid m$ is an infinite set. For any $x \in J(r(H))$, we may write

$$x = d \sum_{[X] \in \operatorname{ind}(\mathcal{C})} \lambda_{[X]}[X],$$

where $d \in \mathbb{Z}$ and all integer coefficients $\lambda_{[X]}$ are coprime. By Lemma 4.1.5 (2) we have $J(r(H)) \subseteq pr(H)$ for all $p \in \Omega$. It follows that $p \mid d$ for all $p \in \Omega$. Thus, $d = 0$, and hence $x = 0$. □

If the Green ring $r(H)$ is commutative, then the Jacobson radical of $r(H)$ is the set of all nilpotent elements of $r(H)$. As a consequence, Theorem 4.1.6 gives a characterization of a commutative Green ring without nonzero nilpotent elements. In particular, if H is a finite group of finite representation type, then the Green ring $r(H)$ is commutative. In this case, the Green ring $r(H)$ has no nonzero nilpotent elements if and only if the Casimir number of $r(H)$ is not zero.

4.2 The Green ring of a cyclic group

In general, it is difficult to calculate the Casimir number of the Green ring $r(H)$. In this section, we focus on the case $H = \mathbb{k}G$, where G is a cyclic group of order p and $\mathbb{k}$ is an algebraically closed field of characteristic p. To describe the Casimir number of the Green ring $r(\mathbb{k}G)$, we first need to describe the structure of the Green ring $r(\mathbb{k}G)$.

From now on p is an odd prime number, $\mathbb{k}$ is an algebraically closed field of characteristic p, and G is a cyclic group of order p. The group algebra $\mathbb{k}G$ is isomorphic to the quotient of the polynomial algebra $\mathbb{k}[X]$ modulo the ideal generated by $X^p - 1$ or $(X-1)^p$:

$$\mathbb{k}G \cong \mathbb{k}[X]/(X^p - 1) \cong \mathbb{k}[X]/((X-1)^p),$$

where the latter is a commutative Nakayama local algebra over $\mathbb{k}$. We denote

$$M_i = \mathbb{k}[X]/((X-1)^i), \quad \text{for } 1 \leqslant i \leqslant p.$$

Then $\{M_1, M_2, \cdots, M_p\}$ forms a complete set of indecomposable $\Bbbk G$-modules up to isomorphism [4, ChV, Section 4]. Here, each M_i is self-dual since M_i is the unique indecomposable module of dimension i up to isomorphism. Note that M_1 is the trivial $\Bbbk G$-module $\Bbbk$.

We follow from [4, ChV, Section 4] and list all almost split sequences of $\Bbbk G$-modules as follows. The almost split sequence ending at the trivial module M_1 is

$$0 \to M_1 \to M_2 \to M_1 \to 0,$$

and the almost split sequence ending at M_i is

$$0 \to M_i \to M_{i+1} \oplus M_{i-1} \to M_i \to 0, \quad \text{for } 1 < i < p.$$

Note that the sequence

$$0 \to M_i \to M_2 \otimes M_i \to M_i \to 0$$

is also an almost split sequence ending at M_i for $1 \leqslant i < p$ (see [4, ChV, Theorem 4.7]). The uniqueness of an almost split sequence shows that

$$M_2 \otimes M_i \cong M_{i+1} \oplus M_{i-1}, \quad \text{for } 1 < i < p.$$

We also have $M_2 \otimes M_p \cong 2M_p$. This leads to the product $[M_2][M_i] = [M_{i-1}] + [M_{i+1}]$ for $1 < i < p$, and $[M_2][M_p] = 2[M_p]$ in the Green ring $r(\Bbbk G)$ of $\Bbbk G$. The product $[M_i][M_j]$ in $r(\Bbbk G)$ can be described as follows:

Lemma 4.2.1 *For $1 \leqslant i, j \leqslant p$, we have*

(1) *If $i + j \leqslant p$, then $[M_i][M_j] = \sum_{t=0}^{\min\{i,j\}-1} [M_{i+j-1-2t}]$.*

(2) *If $i + j \geqslant p + 1$, then $[M_i][M_j] = (i + j - p)[M_p] + \sum_{t=i+j-p}^{\min\{i,j\}-1} [M_{i+j-1-2t}]$.*

Proof This can be proved by induction on $i + j$, or see [86, Proposition 4.2] for a similar result. □

Let $\mathbb{Z}[X_2, \cdots, X_p]$ be a polynomial ring over $\mathbb{Z}$ with variables $X_2, \cdots, X_p$ and I the ideal of $\mathbb{Z}[X_2, \cdots, X_p]$ generated by the following elements:

$$X_2^2 - X_3 - 1, \ X_2X_3 - X_4 - X_2, \ \cdots, \ X_2X_{p-1} - X_p - X_{p-2}, \ X_2X_p - 2X_p.$$

We have

$$r(\Bbbk G) \cong \mathbb{Z}[X_2, \cdots, X_p]/I,$$

where the isomorphism is given by $[M_i] \mapsto \overline{X_i}$ for $i = 2, 3, \cdots, p$ (see [4, ChV, Proposition 4.11]). Actually, the Green ring $r(\Bbbk G)$ is isomorphic to a polynomial ring over $\mathbb{Z}$ with one variable modulo a relation. To see this, we recall the Dickson polynomials of the second kind defined recursively as follows:

$$E_0(X) = 1, \ E_1(X) = X, \ E_{i+1}(X) = XE_i(X) - E_{i-1}(X), \quad \text{for } i \geqslant 1. \tag{4.1}$$

Then $E_n(X)$ can be written explicitly as

$$E_n(X) = \sum_{i=0}^{[\frac{n}{2}]} \binom{n-i}{i} (-1)^i X^{n-2i},$$

for $n \geqslant 0$ (see e.g., [17, Eq.(1.2)]).

Proposition 4.2.2 *We have* $r(\Bbbk G) \cong \mathbb{Z}[X]/((X-2)E_{p-1}(X))$.

Proof Consider the following ring epimorphism

$$\begin{aligned}\psi : \mathbb{Z}[X_2, \cdots, X_p] &\to \mathbb{Z}[X]/((X-2)E_{p-1}(X)),\\ g(X_2, \cdots, X_p) &\mapsto \overline{g(E_1(X), \cdots, E_{p-1}(X))}.\end{aligned}$$

By (4.1) we have $\psi(I) = 0$. This induces a ring epimorphism $\overline{\psi}$ from $\mathbb{Z}[X_2, \cdots, X_p]/I$ to $\mathbb{Z}[X]/((X-2)E_{p-1}(X))$ such that the following diagram is commutative:

$$\begin{array}{ccc} \mathbb{Z}[X_2, \cdots, X_p] & \xrightarrow{\ \psi\ } & \mathbb{Z}[X]/((X-2)E_{p-1}(X)) \\ {\scriptstyle \pi}\downarrow & \nearrow_{\overline{\psi}} & \\ \mathbb{Z}[X_2, \cdots, X_p]/I & & \end{array}$$

where π is the canonical ring epimorphism. Define another ring morphism φ from $\mathbb{Z}[X]$ to $\mathbb{Z}[X_2, \cdots, X_p]/I$ by $\varphi(f(X)) = \overline{f(X_2)}$. By induction on i one is able to check that $\overline{E_{i-1}(X_2)} = \overline{X_i}$ holds in $\mathbb{Z}[X_2, \cdots, X_p]/I$ for $i = 2, 3, \cdots, p$. Thus, φ is surjective. In particular,

$$\varphi((X-2)E_{p-1}(X)) = \overline{(X_2-2)E_{p-1}(X_2)} = \overline{(X_2-2)X_p} = 0.$$

Hence φ induces a ring epimorphism $\overline{\varphi}$ from $\mathbb{Z}[X]/((X-2)E_{p-1}(X))$ to $\mathbb{Z}[X_2, \cdots, X_p]/I$ such that the following diagram is commutative:

$$\begin{array}{ccc} \mathbb{Z}[X] & \xrightarrow{\ \varphi\ } & \mathbb{Z}[X_2, \cdots, X_p]/I \\ {\scriptstyle \pi}\downarrow & \nearrow_{\overline{\varphi}} & \\ \mathbb{Z}[X]/((X-2)E_{p-1}(X)) & & \end{array}$$

Now it is straightforward to check that $\overline{\psi} \circ \overline{\varphi} = id$ and $\overline{\varphi} \circ \overline{\psi} = id$, as desired. □

The almost split sequences of $\Bbbk G$-modules are useful to calculate dimensions of morphism spaces. We illustrate it here, although it is not closely related to the topic of this section. According to the notation of $\delta_{[M]}$, we have

$$\delta_{[M_i]} = \begin{cases} 2 - [M_2], & i = 1, \\ 2[M_i] - [M_{i+1}] - [M_{i-1}], & 1 < i < p, \\ [M_p] - [M_{p-1}], & i = p. \end{cases}$$

In particular, we have $\delta_{[M_i]} = \delta_{[M_1]}[M_i]$ for $1 \leqslant i < p$ and $\delta_{[M_1]}[M_p] = 0$. This gives the following relationship between the bases $\{\delta_{[M_i]} \mid 1 \leqslant i \leqslant p\}$ and $\{[M_i] \mid 1 \leqslant i \leqslant p\}$ of $r(\Bbbk G)$:

$$\begin{pmatrix} \delta_{[M_1]} \\ \delta_{[M_2]} \\ \vdots \\ \delta_{[M_{p-1}]} \\ \delta_{[M_p]} \end{pmatrix} = \begin{pmatrix} 2 & -1 & & & \\ -1 & 2 & -1 & & \\ & \ddots & \ddots & \ddots & \\ & & -1 & 2 & -1 \\ & & & -1 & 1 \end{pmatrix} \begin{pmatrix} [M_1] \\ [M_2] \\ \vdots \\ [M_{p-1}] \\ [M_p] \end{pmatrix}.$$

Note that

$$\begin{pmatrix} 2 & -1 & & & \\ -1 & 2 & -1 & & \\ & \ddots & \ddots & \ddots & \\ & & -1 & 2 & -1 \\ & & & -1 & 1 \end{pmatrix}^{-1} = \begin{pmatrix} 1 & 1 & \cdots & 1 & 1 \\ 1 & 2 & \cdots & 2 & 2 \\ \vdots & \vdots & \ddots & \vdots & \vdots \\ 1 & 2 & \cdots & p-1 & p-1 \\ 1 & 2 & \cdots & p-1 & p \end{pmatrix}$$

whose (i,j)-entry is $\min\{i,j\}$. We have

$$[M_i] = \sum_{j=1}^{p} ([M_i],[M_j])\delta^*_{[M_j]} = \sum_{j=1}^{p} \dim_{\Bbbk} \mathrm{Hom}_{\Bbbk G}(M_i, M_j)\delta_{[M_j]},$$

since the duality operator $*$ on $r(\Bbbk G)$ is the identity map. It follows that

$$\dim_{\Bbbk} \mathrm{Hom}_{\Bbbk G}(M_i, M_j) = \min\{i,j\}.$$

4.3 The Casimir number of the Green ring of a cyclic group

In this section, $\Bbbk$ is an algebraically closed field of characteristic p, and G is a cyclic group of order p. We shall compute the Casimir number of the Green ring $r(\Bbbk G)$. Using this number we could explicitly determine the Jacobson radical of the Green algebra $r(\Bbbk G) \otimes_{\mathbb{Z}} K$ for any field K.

The Casimir operator c of $r(\Bbbk G)$ is given by

$$c(x) = \sum_{i=1}^{p} [M_i]x\delta^*_{[M_i]} = \sum_{i=1}^{p} [M_i]x\delta_{[M_i]} = xc(1), \quad \text{for } x \in r(\Bbbk G)$$

since $r(\Bbbk G)$ is commutative. To determine the Casimir number of $r(\Bbbk G)$, we need to describe the Casimir element $c(1)$ of $r(\Bbbk G)$.

Lemma 4.3.1 *We have* $c(1) = [M_p] + 2\sum_{i=1}^{p}(-1)^{i-1}(p-i)[M_i]$.

Proof Firstly, it is straightforward to check by Lemma 4.2.1 that

$$\sum_{i=1}^{\frac{p-1}{2}}[M_i]^2 = \sum_{i=1}^{\frac{p-1}{2}}(\frac{p+1}{2} - i)[M_{2i-1}] = \sum_{i=\frac{p+1}{2}}^{p-1}[M_i]^2.$$

Using this equality we have

$$\begin{aligned}
c(1) &= \sum_{i=1}^{p}[M_i]\delta_{[M_i]} = \delta_{[M_1]}\sum_{i=1}^{p-1}[M_i]^2 + [M_p]([M_p] - [M_{p-1}]) \\
&= [M_p] + \delta_{[M_1]}(\sum_{i=1}^{\frac{p-1}{2}}[M_i]^2 + \sum_{i=\frac{p+1}{2}}^{p-1}[M_i]^2) \\
&= [M_p] + 2\delta_{[M_1]}\sum_{i=1}^{\frac{p-1}{2}}[M_i]^2 \\
&= [M_p] + 2\delta_{[M_1]}\sum_{i=1}^{\frac{p-1}{2}}(\frac{p+1}{2} - i)[M_{2i-1}] \\
&= [M_p] + 2(2 - [M_2])\sum_{i=1}^{\frac{p-1}{2}}(\frac{p+1}{2} - i)[M_{2i-1}] \\
&= [M_p] + 2\sum_{i=1}^{\frac{p-1}{2}}(\frac{p+1}{2} - i)(2[M_{2i-1}] - [M_{2i}] - [M_{2i-2}]) \\
&= [M_p] + 2\sum_{i=1}^{p}(-1)^{i-1}(p-i)[M_i].
\end{aligned}$$

The proof is completed. □

Since the duality operator $*$ on $r(\Bbbk G)$ is the identity map, it can be seen from Remark 1.4.4 that $\{\delta_{[M_t]}, [M_t] \mid t = 1, 2, \cdots, p\}$ forms a pair of dual bases of $r(\Bbbk G)$ with respect to the bilinear form $(-,-)$ given in (1.11). For any $x \in r(\Bbbk G)$, by Lemma 1.4.2 (2) we have

$$c(x) = \sum_{t=1}^{p}(\delta_{[M_t]}, c(x))[M_t].$$

Thus, the coefficient of $[M_t]$ in the linear expression of $c(x)$ is $(\delta_{[M_t]}, c(x))$. Next, we compute the value of $(\delta_{[M_t]}, c(x))$ for each $1 \leqslant t \leqslant p$.

Lemma 4.3.2 *If* $x = \sum_{j=1}^{p} \lambda_j [M_j]$, *then* $(\delta_{[M_p]}, c(x)) = \sum_{i=1}^{\frac{p+1}{2}} (2i-1)\lambda_{2i-1}$.
Proof By Lemma 4.3.1, we have

$$\begin{aligned}
c(x) = c(1)x &= ([M_p] + 2\sum_{i=1}^{p}(-1)^{i-1}(p-i)[M_i])\sum_{j=1}^{p}\lambda_j[M_j] \\
&= \sum_{j=1}^{p} j\lambda_j[M_p] + 2\sum_{i,j=1}^{p}(-1)^{i-1}(p-i)\lambda_j[M_i][M_j] \\
&= \sum_{j=1}^{p} j\lambda_j[M_p] + 2\sum_{i+j=p+1}^{2p}(-1)^{i-1}(p-i)(i+j-p)\lambda_j[M_p] + \sum_{i=1}^{p-1}\mu_i[M_i],
\end{aligned}$$

where the last equality follows from Lemma 4.2.1 (2) with some $\mu_i \in \mathbb{Z}$. Then

$$\begin{aligned}
(\delta_{[M_p]}, c(x)) &= \sum_{j=1}^{p} j\lambda_j + 2\sum_{i+j=p+1}^{2p}(-1)^{i-1}(p-i)(i+j-p)\lambda_j \\
&= \sum_{j=1}^{p} j\lambda_j + 2\sum_{j=1}^{p}\sum_{i=p+1-j}^{p}(-1)^{i-1}(p-i)(i+j-p)\lambda_j \\
&= \sum_{j=1}^{p} j\lambda_j + 2\sum_{j=1}^{p}\sum_{k=1}^{j}(-1)^{k-j}(j-k)k\lambda_j.
\end{aligned}$$

Note that

$$\sum_{k=1}^{j}(-1)^k(j-k)k = \begin{cases} 0, & 2 \nmid j, \\ -\dfrac{j}{2}, & 2 \mid j. \end{cases}$$

Thus,

$$(\delta_{[M_p]}, c(x)) = \sum_{j=1}^{p} j\lambda_j - \sum_{2|j,j=1}^{p} j\lambda_j = \sum_{2\nmid j,j=1}^{p} j\lambda_j = \sum_{i=1}^{\frac{p+1}{2}}(2i-1)\lambda_{2i-1}.$$

The proof is completed. □

To describe the value of $(\delta_{[M_t]}, c(x))$ for each $1 \leqslant t \leqslant p-1$, we need some preparations. The left multiplication by $[M_t]$ with respect to the basis $\{[M_1], [M_2], \cdots, [M_p]\}$ corresponds to a matrix $\mathbf{M}_t$. That is,

$$[M_t]\begin{pmatrix} [M_1] \\ [M_2] \\ \vdots \\ [M_p] \end{pmatrix} = \mathbf{M}_t \begin{pmatrix} [M_1] \\ [M_2] \\ \vdots \\ [M_p] \end{pmatrix}.$$

If we denote by $\mathbf{E}_{i,j}$ the square matrix of order p with (i,j)-entry 1, and 0 otherwise, then $\mathbf{M}_t$ can be written explicitly as follows:

$$\begin{aligned}\mathbf{M}_t &= \mathbf{E}_{1,t} + \mathbf{E}_{2,t+1} + \mathbf{E}_{3,t+2} + \cdots + \mathbf{E}_{p-t,p-1} \\ &\quad + \mathbf{E}_{2,t-1} + \mathbf{E}_{3,t} + \mathbf{E}_{4,t+1} + \cdots + \mathbf{E}_{p-t+1,p-2} \\ &\quad + \mathbf{E}_{3,t-2} + \mathbf{E}_{4,t-1} + \mathbf{E}_{5,t} + \cdots + \mathbf{E}_{p-t+2,p-3} \\ &\quad + \cdots \\ &\quad + \mathbf{E}_{t,1} + \mathbf{E}_{t+1,2} + \mathbf{E}_{t+2,3} + \cdots + \mathbf{E}_{p-1,p-t} \\ &\quad + t\mathbf{E}_{p,p} + (t-1)\mathbf{E}_{p-1,p} + (t-2)\mathbf{E}_{p-2,p} + \cdots + \mathbf{E}_{p-t+1,p} \\ &= \sum_{s=1}^{t}\sum_{r=1}^{p-t} \mathbf{E}_{s+r-1,t+r-s} + \sum_{s=1}^{t}(t+1-s)\mathbf{E}_{p-s+1,p}. \end{aligned} \tag{4.2}$$

Lemma 4.3.3 *If $x = \sum_{j=1}^{p} \lambda_j [M_j]$, then*

$$(\delta_{[M_t]}, c(x)) = 2(p-t)\sum_{i=1}^{t}(-1)^{t+i} i\lambda_i + 2t\sum_{i=t+1}^{p-1}(-1)^{t+i}(p-i)\lambda_i,$$

for $1 \leqslant t \leqslant p-1$.

Proof Let

$$[M_i][M_j] = \sum_{t=1}^{p} N_{ij}^t [M_t], \quad \text{for } N_{ij}^t \in \mathbb{Z}.$$

For $1 \leqslant i,j,t \leqslant p-1$, the associativity of the form $(-,-)$ over $r(\Bbbk G)$ together with the commutativity of $r(\Bbbk G)$ shows that

$$\begin{aligned} N_{ij}^t &= (\delta_{[M_t]}, [M_i][M_j]) = (\delta_{[M_1]}[M_t], [M_i][M_j]) \\ &= (\delta_{[M_1]}[M_j], [M_t][M_i]) = (\delta_{[M_j]}, [M_t][M_i]) \\ &= N_{ti}^j. \end{aligned} \tag{4.3}$$

Consequently, we have

$$\begin{aligned} c(x) = c(1)x &= \left([M_p] + 2\sum_{i=1}^{p}(-1)^{i-1}(p-i)[M_i]\right)\sum_{j=1}^{p}\lambda_j[M_j] \\ &= \mu_1[M_p] + 2\sum_{i,j=1}^{p-1}(-1)^{i-1}(p-i)\lambda_j[M_i][M_j] \quad (\text{for some } \mu_1 \in \mathbb{Z}) \end{aligned}$$

$$= \mu_1[M_p] + 2\sum_{i,j=1}^{p-1}(-1)^{i-1}(p-i)\lambda_j\sum_{t=1}^{p}N_{ij}^t[M_t]$$

$$= \mu_2[M_p] + 2\sum_{i,j,t=1}^{p-1}(-1)^{i-1}(p-i)\lambda_j N_{ij}^t[M_t] \quad (\text{for some } \mu_2 \in \mathbb{Z})$$

$$= \mu_2[M_p] + 2\sum_{i,j,t=1}^{p-1}(-1)^{i-1}(p-i)\lambda_j N_{ti}^j[M_t] \quad (\text{by } (4.3)).$$

Thus,

$$(\delta_{[M_t]}, c(x)) = 2\sum_{i,j=1}^{p-1}(-1)^{i-1}(p-i)\lambda_j N_{ti}^j, \quad \text{for } 1 \leqslant t \leqslant p-1.$$

Let $\widehat{\mathbf{M}}_t$ be the submatrix of $\mathbf{M}_t$ obtained by deleting the p-th column and row. By (4.2) we have

$$\widehat{\mathbf{M}}_t = \sum_{s=1}^{t}\sum_{r=1}^{p-t}\mathbf{E}_{s+r-1,t+r-s}.$$

The matrix $\widehat{\mathbf{M}}_t$ is symmetric and

$$\begin{aligned}
(\delta_{[M_t]}, c(x)) &= 2\sum_{i,j=1}^{p-1}(-1)^{i-1}(p-i)\lambda_j N_{ti}^j \\
&= 2\begin{pmatrix}\lambda_1 & \lambda_2 & \cdots & \lambda_{p-1}\end{pmatrix}\widehat{\mathbf{M}}_t\begin{pmatrix} p-1 \\ -(p-2) \\ \vdots \\ (-1)^{p-2}\end{pmatrix} \\
&= 2\begin{pmatrix}\lambda_1 & \lambda_2 & \cdots & \lambda_{p-1}\end{pmatrix}\sum_{s=1}^{t}\sum_{r=1}^{p-t}\mathbf{E}_{s+r-1,t+r-s}\begin{pmatrix} p-1 \\ -(p-2) \\ \vdots \\ (-1)^{p-2}\end{pmatrix} \\
&= 2\sum_{s=1}^{t}\sum_{r=1}^{p-t}(-1)^{t+r-s-1}(p-t-r+s)\lambda_{s+r-1} \\
&= 2(p-t)\sum_{i=1}^{t}(-1)^{t+i}i\lambda_i + 2t\sum_{i=t+1}^{p-1}(-1)^{t+i}(p-i)\lambda_i.
\end{aligned}$$

We are done. □

The Casimir number of $r(\Bbbk G)$ can be presented as follows:

Theorem 4.3.4 *The Casimir number of $r(\Bbbk G)$ is $2p^2$.*

Proof Let $x = \sum_{j=1}^{p} \lambda_j [M_j]$. Then $c(x) = \sum_{t=1}^{p} (\delta_{[M_t]}, c(x))[M_t]$. If $c(x) \in \mathbb{Z}$, then $(\delta_{[M_t]}, c(x)) = 0$ for $t = 2, 3, \cdots, p$. However,

$$(\delta_{[M_t]}, c(x)) = 2(p-t)\sum_{i=1}^{t}(-1)^{t+i} i\lambda_i + 2t\sum_{i=t+1}^{p-1}(-1)^{t+i}(p-i)\lambda_i, \tag{4.4}$$

for $t = 2, 3, \cdots, p-1$ (see Lemma 4.3.3), and

$$(\delta_{[M_p]}, c(x)) = \sum_{i=1}^{\frac{p+1}{2}} (2i-1)\lambda_{2i-1} \quad \text{(see Lemma 4.3.2).}$$

This gives a system of equations with variables $\lambda_1, \cdots, \lambda_p$. Consider the following equations:

$$\begin{cases} (\delta_{[M_{p-1}]}, c(x)) = 0, \\ (\delta_{[M_{p-2}]}, c(x)) = 0. \end{cases}$$

Using (4.4) it is not hard to see that $\lambda_{p-1} = 0$. Similarly, the system of equations

$$\begin{cases} (\delta_{[M_{p-2}]}, c(x)) = 0, \\ (\delta_{[M_{p-3}]}, c(x)) = 0 \end{cases}$$

together with $\lambda_{p-1} = 0$ shows that $\lambda_{p-2} = 0$. Repeating this argument we obtain that $\lambda_{p-1} = \lambda_{p-2} = \cdots = \lambda_3 = 0$. Now the system of equations

$$\begin{cases} (\delta_{[M_2]}, c(x)) = 0, \\ (\delta_{[M_p]}, c(x)) = 0 \end{cases}$$

can be simplified as follows:

$$\begin{cases} -\lambda_1 + 2\lambda_2 = 0, \\ \lambda_1 + p\lambda_p = 0. \end{cases}$$

It follows that $\lambda_1 = 2p\mu$, $\lambda_2 = p\mu$, $\lambda_p = -2\mu$ for $\mu \in \mathbb{Z}$. In this case,

$$\begin{aligned}(\delta_{[M_1]}, c(x)) &= 2(p-1)\lambda_1 + 2\sum_{i=2}^{p-1}(-1)^{1+i}(p-i)\lambda_i \\ &= 2(p-1)\lambda_1 - 2(p-2)\lambda_2 \\ &= 2p^2\mu.\end{aligned}$$

We conclude that $\mathrm{Imc} \cap \mathbb{Z} = (2p^2)$. □

Since the Casimir number of $r(\Bbbk G)$ is $2p^2 \neq 0$, the Green ring $r(\Bbbk G)$ is semisimple. This is exactly a result of [34]. For the Green algebra $r(\Bbbk G) \otimes_{\mathbb{Z}} K$, it follows from Theorem 4.1.2 that $r(\Bbbk G) \otimes_{\mathbb{Z}} K$ is semisimple if and only if the characteristic of K is not equal to 2 or p. Next, we use the factorizations of Dickson polynomials to determine the generators of the Jacobson radical of $r(\Bbbk G) \otimes_{\mathbb{Z}} K$, or equivalently, $K[X]/((X-2)E_{p-1}(X))$ (see Proposition 4.2.2) in the cases where K is of characteristic 2 or p.

Proposition 4.3.5 *If the characteristic of K is p, then the Jacobson radical of the Green algebra $K[X]/((X-2)E_{p-1}(X))$ is a principal ideal generated by $\overline{X^2-4}$.*

Proof We have the decomposition

$$E_{p-1}(X) = (X-2)^{\frac{p-1}{2}}(X+2)^{\frac{p-1}{2}}$$

in $K[X]$ [17, Theorem 3.1 (2)]. Thus, the polynomial $(X-2)E_{p-1}(X)$ has only two distinct prime factors $X-2$ and $X+2$. Since $K[X]$ is a principal ideal domain and every nonzero prime ideal is maximal, the Jacobson radical of $K[X]/((X-2)E_{p-1}(X))$ is a principal ideal generated by $\overline{(X-2)(X+2)}$, which is the product of distinct prime factors of $(X-2)E_{p-1}(X)$. □

Proposition 4.3.6 *If the characteristic of K is 2, then the Jacobson radical of the Green algebra $K[X]/((X-2)E_{p-1}(X))$ is a principal ideal generated by*

$$\overline{\sum_{i=0}^{[\frac{p-1}{2}]} \binom{p-1-i}{i} (-1)^i X^{\frac{p+1}{2}-i}}.$$

Proof Since the characteristic of K is 2, we have the following isomorphism:

$$K[X]/((X-2)E_{p-1}(X)) \cong K[X]/(XE_{p-1}(X)).$$

The Dickson polynomial $E_{p-1}(X)$ in $K[X]$ can be written as

$$E_{p-1}(X) = \sum_{i=0}^{[\frac{p-1}{2}]} \binom{p-1-i}{i} (-1)^i X^{p-1-2i} = (f(X))^2,$$

where

$$f(X) = \sum_{i=0}^{[\frac{p-1}{2}]} \binom{p-1-i}{i} (-1)^i X^{\frac{p-1}{2}-i},$$

and it has no multiple factors in $K[X]$, see [9, Theorem 6]. It follows that the Jacobson radical of $K[X]/(XE_{p-1}(X))$ is a principal ideal generated by $\overline{Xf(X)}$. The proof is completed. □

Chapter 5 The Casimir Numbers of Fusion Categories

Throughout this chapter, unless otherwise stated, all fusion categories and Hopf algebras are defined over an algebraically closed field $\Bbbk$ of arbitrary characteristic. In this chapter, we associate any fusion category with the Casimir number. The Casimir number of a fusion category plays a role similar to the Casimir number of the Green ring of a Hopf algebra. Applying to the Verlinde modular category of rank $n+1$, we find that its Casimir number is $2n+4$. We then give some results concerning prime factors of the Casimir number of a fusion category. We reveal a relationship between the Casimir number of a fusion category $\mathcal{C}$ and the Casimir number of the pivotalization of $\mathcal{C}$. We show that the former is a factor of the latter. This gives a result that any non-degenerate fusion category has a nonzero Casimir number. We finally prove that the Casimir number and the Frobenius-Schur exponent of a spherical fusion category have the same prime factors. This may be considered as an integer version of the Cauchy theorem for a spherical fusion category.

5.1 Numerical invariants

In this section, we introduce two numerical invariants of a fusion category $\mathcal{C}$, one of them is called the Casimir number of $\mathcal{C}$. We use the Casimir number to determine when the Grothendieck algebra $\mathrm{Gr}(\mathcal{C})\otimes_{\mathbb{Z}} K$ over a field K is semisimple. We also use the Casimir number to determine when a pivotal fusion category is non-degenerate.

Let $\mathcal{C}$ be a fusion category over a field $\Bbbk$ and $\{X_i\}_{i\in I}$ the set of isomorphism classes of simple objects of $\mathcal{C}$. The *Grothendieck ring* $\mathrm{Gr}(\mathcal{C})$ of $\mathcal{C}$ is an associative unital ring with a multiplication induced by the tensor product on $\mathcal{C}$, namely,

$$X_iX_j = \sum_{k\in I} N_{ij}^k X_k,$$

where N_{ij}^k, called the *fusion coefficient* of $\mathrm{Gr}(\mathcal{C})$, is the multiplicity of X_k in the Jordan-Hölder series of $X_i \otimes X_j$. The duality functor $*$ of $\mathcal{C}$ induces an involution on $\mathrm{Gr}(\mathcal{C})$, namely,

$$(X_iX_j)^* = X_j^*X_i^* \quad \text{and} \quad (X_i)^{**} = X_i, \quad \text{for } i, j \in I.$$

We write $(X_i)^* = X_{i^*}$ for convenience. In view of this, the duality functor $*$ induces a permutation on the index set I.

There is an associative symmetric and non-degenerate $\mathbb{Z}$-bilinear form $(-,-)$ on $\mathrm{Gr}(\mathcal{C})$ defined by

$$(X_i, X_j) = \dim_{\Bbbk} \mathrm{Hom}(X_i, X_j^*) = \delta_{i,j^*},$$

where δ_{i,j^*} is the Kronecker symbol. This form is also $*$-invariant, namely,

$$(X_i, X_j) = (X_i^*, X_j^*), \quad \text{for all } i, j \in I.$$

Thus, $\mathrm{Gr}(\mathcal{C})$ is a symmetric $*$-algebra over $\mathbb{Z}$. The pair of dual bases with respect to the form $(-,-)$ is the set $\{X_i, X_{i^*}\}_{i\in I}$ satisfying the following equality:

$$\sum_{i\in I} X_i \otimes X_{i^*} = \sum_{i\in I} X_{i^*} \otimes X_i.$$

Note that $N_{ij}^k = (X_iX_j, X_{k^*})$ for all $i, j, k \in I$. It follows from

$$(X_iX_j, X_{k^*}) = (X_{i^*}X_k, X_{j^*}) = (X_kX_{j^*}, X_{i^*})$$

that $N_{ij}^k = N_{i^*k}^j = N_{kj^*}^i$. Using this equality one is able to check the following two equalities:

$$\sum_{i\in I} X_jX_i \otimes X_{i^*} = \sum_{i\in I} X_i \otimes X_{i^*}X_j, \tag{5.1}$$

$$\sum_{i\in I} X_iX_j \otimes X_{i^*} = \sum_{i\in I} X_i \otimes X_jX_{i^*}. \tag{5.2}$$

The *Casimir operator* (see e.g., [52, Section 3.1]) of the Grothendieck ring $\mathrm{Gr}(\mathcal{C})$ is the map c from $\mathrm{Gr}(\mathcal{C})$ to its center $Z(\mathrm{Gr}(\mathcal{C}))$ given by

$$c(a) = \sum_{i\in I} X_iaX_{i^*}, \quad \text{for } a \in \mathrm{Gr}(\mathcal{C}).$$

The element $c(1) = \sum_{i\in I} X_iX_{i^*}$, depending on $(-,-)$ only up to a central unit of $\mathrm{Gr}(\mathcal{C})$ (see [52, Section 1.2.5]), is called the *Casimir element* of $\mathrm{Gr}(\mathcal{C})$.

If we consider $c(1) = \sum_{i\in I} X_iX_{i^*}$ as an element in $\mathrm{Gr}(\mathcal{C}) \otimes_{\mathbb{Z}} \mathbb{Q}$, then it is central invertible (see the proof of [28, Lemma 9.3.10]). Hence there exists a unique central invertible element $b \in \mathrm{Gr}(\mathcal{C}) \otimes_{\mathbb{Z}} \mathbb{Q}$ such that $c(1)b = 1$. Suppose $b = \sum_{i\in I} \frac{m_i}{n_i} X_i$, where m_i and n_i form a pair of coprime integers for each $i \in I$. Denote by $n_{\mathcal{C}} > 0$ the least common multiple of n_i for all $i \in I$. Then $bn_{\mathcal{C}} \in \mathrm{Gr}(\mathcal{C})$ and $n_{\mathcal{C}} = c(1)bn_{\mathcal{C}} = c(bn_{\mathcal{C}})$. This means that $n_{\mathcal{C}} \in \mathbb{Z} \cap \mathrm{Im}c$, and hence $\mathbb{Z} \cap \mathrm{Im}c \neq (0)$.

Since the intersection $\mathbb{Z} \cap \mathrm{Im} c$ is a nonzero principle ideal of $\mathbb{Z}$, the positive generator of $\mathbb{Z} \cap \mathrm{Im} c$, denoted by $m_{\mathcal{C}}$, is called the *Casimir number* of $\mathcal{C}$. Namely, $\mathbb{Z} \cap \mathrm{Im} c = (m_{\mathcal{C}})$ for $m_{\mathcal{C}} > 0$. The element a satisfying $c(a) = m_{\mathcal{C}}$ is not unique in general. It is easy to see that the element a satisfying $c(a) = m_{\mathcal{C}}$ is unique if and only if the map c is injective, if and only if $\mathrm{Gr}(\mathcal{C})$ is commutative. The Casimir number $m_{\mathcal{C}}$ always divides the number $n_{\mathcal{C}}$ since we have seen that $n_{\mathcal{C}} \in \mathbb{Z} \cap \mathrm{Im} c$. If $\mathrm{Gr}(\mathcal{C})$ is commutative, we have $m_{\mathcal{C}} = n_{\mathcal{C}}$.

Observe that the matrix $[c(1)]$ of left multiplication by $c(1)$ with respect to the basis $\{X_i\}_{i \in I}$ of $\mathrm{Gr}(\mathcal{C})$ is a positive definite integer matrix (see [52, Proposition 8]). It follows that the determinant $d_{\mathcal{C}} := \det[c(1)]$, called the *determinant* of $\mathcal{C}$, is always a positive integer.

Remark 5.1.1 (1) *If two fusion categories are monoidally equivalent under a monoidal functor, then this functor induces an isomorphism preserving fusion coefficients between the Grothendieck rings of fusion categories. Thus, monoidally equivalent fusion categories lead to the same Casimir numbers and the same determinants. In other words, the Casimir number $m_{\mathcal{C}}$ and the determinant $d_{\mathcal{C}}$ are both monoidal invariants of a fusion category $\mathcal{C}$.*

(2) *Suppose the characteristic polynomial of the integer matrix $[c(1)]$ is*

$$f(x) = x^n + \alpha_1 x^{n-1} + \cdots + \alpha_{n-1} x + \alpha_n,$$

where n is the cardinality of I, $f(x) \in \mathbb{Z}[x]$ and $\alpha_n = \pm d_{\mathcal{C}}$. By the Cayley-Hamilton's theorem, the operator of left multiplication by $c(1)$ satisfies that

$$0 = f(c(1)) = c(1)(c(1)^{n-1} + \alpha_1 c(1)^{n-2} + \cdots + \alpha_{n-1}) + \alpha_n = c(1)a + \alpha_n,$$

where

$$a = c(1)^{n-1} + \alpha_1 c(1)^{n-2} + \cdots + \alpha_{n-1} \in Z(\mathrm{Gr}(\mathcal{C})).$$

Thus,

$$c(a) = c(1)a = -\alpha_n = \mp d_{\mathcal{C}} \in \mathbb{Z}.$$

By the definition of $m_{\mathcal{C}}$, we have $m_{\mathcal{C}} \mid d_{\mathcal{C}}$.

(3) *The Casimir number of a fusion category defined here is indeed a special case of the Casimir number of the Green ring of a finite dimensional Hopf algebra (see Section 4.1).*

Proposition 5.1.2 *Let $\mathcal{C}$ be a fusion category over $\Bbbk$. For any field K, the Casimir number $m_{\mathcal{C}} \neq 0$ in K if and only if the Grothendieck algebra $\mathrm{Gr}(\mathcal{C}) \otimes_{\mathbb{Z}} K$ is semisimple.*

Proof The result is essentially the theorem 4.1.2 applied to the Grothendieck algebra $\mathrm{Gr}(\mathcal{C}) \otimes_{\mathbb{Z}} K$. □

A fusion category $\mathcal{C}$ over a field $\Bbbk$ is called *non-degenerate* if the global dimension $\dim(\mathcal{C})$ of $\mathcal{C}$ is not zero in $\Bbbk$. Recall from [73, Theorem 6.5] that a pivotal fusion category $\mathcal{C}$ over a field $\Bbbk$ is non-degenerate if and only if its Grothendieck algebra $\mathrm{Gr}(\mathcal{C}) \otimes_{\mathbb{Z}} \Bbbk$ is semisimple. This result together with Proposition 5.1.2 gives the following numerical characterization of a non-degenerate pivotal fusion category:

Proposition 5.1.3 *A pivotal fusion category $\mathcal{C}$ over a field $\Bbbk$ is non-degenerate if and only if the Casimir number $m_{\mathcal{C}}$ is not zero in $\Bbbk$.*

Proposition 5.1.4 *Let $\mathcal{C}$ be a fusion category over a field $\Bbbk$ with a commutative Grothendieck ring $\mathrm{Gr}(\mathcal{C})$. For any field K, the following statements are equivalent:*

(1) *The determinant $d_{\mathcal{C}} \neq 0$ in K.*

(2) *The Casimir number $m_{\mathcal{C}} \neq 0$ in K.*

(3) *The Grothendieck algebra $\mathrm{Gr}(\mathcal{C}) \otimes_{\mathbb{Z}} K$ is semisimple.*

Proof (1) $\Rightarrow$ (2). It follows from Remark 5.1.1 (2) that $m_{\mathcal{C}} \mid d_{\mathcal{C}}$. Thus, if $d_{\mathcal{C}} \neq 0$ in K, then $m_{\mathcal{C}} \neq 0$ in K.

(2) $\Rightarrow$ (3). Note that there exists some $a \in \mathrm{Gr}(\mathcal{C})$ such that

$$\sum_{i \in I} X_i a X_{i^*} = m_{\mathcal{C}}.$$

Denote $A := \mathrm{Gr}(\mathcal{C}) \otimes_{\mathbb{Z}} K$ and consider $\sum_{i \in I} X_i \frac{a}{m_{\mathcal{C}}} \otimes X_{i^*} \in A \otimes A$. Obviously,

$$\sum_{i \in I} X_i \frac{a}{m_{\mathcal{C}}} X_{i^*} = 1$$

and

$$\sum_{i \in I} b X_i \frac{a}{m_{\mathcal{C}}} \otimes X_{i^*} = \sum_{i \in I} X_i \frac{a}{m_{\mathcal{C}}} \otimes X_{i^*} b$$

hold for any $b \in A$, see (5.1). Thus,

$$\sum_{i \in I} X_i \frac{a}{m_{\mathcal{C}}} \otimes X_{i^*}$$

is a separable idempotent of A, and hence A is a separable K-algebra. It is well known that any separable K-algebra is a semisimple K-algebra (see e.g., [11]).

(3) $\Rightarrow$ (1). Let $\mathrm{Tr}(a)$ be the trace of the operator of left multiplication by $a \in A = \mathrm{Gr}(\mathcal{C}) \otimes_{\mathbb{Z}} K$. Since A is commutative semisimple, the bilinear form $\langle a, b \rangle = \mathrm{Tr}(ab)$ on A is non-degenerate. This implies that the matrix $[a_{ij}]$ for $a_{ij} = \langle X_i, X_j \rangle$ is an invertible matrix in K. Let c_{ij} be the (i,j)-entry of $[c(1)]$. Then

$$\begin{aligned} c_{ij} &= (\sum_{k \in I} X_k X_{k^*} X_i, X_{j^*}) = \sum_{k \in I} (X_k X_{k^*}, X_i X_{j^*}) = \sum_{k \in I} (X_i X_{j^*}, X_k X_{k^*}) \\ &= \sum_{k \in I} (X_i X_{j^*} X_k, X_{k^*}) = \mathrm{Tr}(X_i X_{j^*}) = a_{ij^*}. \end{aligned}$$

That is, the matrix $[c(1)]$ differs from the matrix $[a_{ij}]$ only by permutations of columns. It follows that $[c(1)]$ is an invertible matrix in K and $\det[c(1)] = d_{\mathcal{C}} \neq 0$ in K. □

Remark 5.1.5 (1) *The proof of* (3) ⇒ (1) *in Proposition* 5.1.4 *comes from the proof of* [68, *Proposition* 2.9]. *From this proof one is able to see that* $d_{\mathcal{C}} = \pm\det[a_{ij}]$, *where* $a_{ij} = \mathrm{Tr}(X_iX_j)$ *for* $i, j \in I$.

(2) *Any one of the statements of Proposition* 5.1.4 *is equivalent to the result that* $c(1) = \sum_{i\in I} X_iX_{i^*}$ *is invertible in* $\mathrm{Gr}(\mathcal{C}) \otimes_{\mathbb{Z}} K$ (*see* [76, *Theorem* 3.8]).

If $\mathrm{char}(K) = p > 0$, it follows from Proposition 5.1.4 that $p \nmid d_{\mathcal{C}}$ if and only if $p \nmid m_{\mathcal{C}}$. This gives the following relationship between the numbers $m_{\mathcal{C}}$ and $d_{\mathcal{C}}$:

Corollary 5.1.6 *The Casimir number* $m_{\mathcal{C}}$ *and the determinant* $d_{\mathcal{C}}$ *of a fusion category* $\mathcal{C}$ *have the same prime factors, if the Grothendieck ring* $\mathrm{Gr}(\mathcal{C})$ *is commutative.*

The rest of this section provides some fusion categories whose determinants or Casimir numbers can be explicitly described.

Example 5.1.7 Let $\mathcal{C}$ be a pointed fusion category over a field $\Bbbk$. The Grothendieck ring of $\mathcal{C}$ is the group ring $\mathbb{Z}G$ for a finite group G. The Casimir number of $\mathcal{C}$ is the order $|G|$ of G and the determinant of $\mathcal{C}$ is $|G|^{|G|}$. It follows from Proposition 5.1.2 that for any field K, the K-algebra $KG = \mathbb{Z}G \otimes_{\mathbb{Z}} K$ is semisimple if and only if $|G| \neq 0$ in K. This is the Maschke's theorem for group algebras.

Example 5.1.8 Let $\mathcal{C}$ be a *modular category* over a field $\Bbbk$ with isomorphism classes of simple objects $\{X_i\}_{i\in I}$. That is, $\mathcal{C}$ is a spherical fusion category with a braiding c such that the S-matrix $\mathbf{S} = [s_{ij}]$ is invertible in $\Bbbk$, where $s_{ij} = \mathrm{Tr}(c_{X_jX_i} \circ c_{X_iX_j})$ (see e.g., [28, Section 8.14]). Note that $\dim(X_i) \neq 0$ in $\Bbbk$ for any $i \in I$ (see [28, Proposition 4.8.4]). For any $i \in I$, the map

$$h_i : X_j \mapsto \frac{s_{ij}}{\dim(X_i)}, \quad \text{for } j \in I$$

defines a homomorphism from $\mathrm{Gr}(\mathcal{C})$ to $\Bbbk$. In other words, $\{h_i(X_j)\}_{i\in I}$ consists of all eigenvalues of the matrix $[X_j]$ of left multiplication by X_j with respect to the basis $\{X_i\}_{i\in I}$ of $\mathrm{Gr}(\mathcal{C})$. Note that all eigenvalues of the matrix $[c(1)]$ are $h_i(c(1))$ for $i \in I$. Moreover,

$$h_i(c(1)) = h_i\left(\sum_{j\in I} X_jX_{j^*}\right) = \sum_{j\in I} h_i(X_j)h_i(X_{j^*}) = \sum_{j\in I} \frac{s_{ij}s_{ij^*}}{\dim(X_i)^2} = \frac{\dim(\mathcal{C})}{\dim(X_i)^2},$$

where the last equality follows from [28, Proposition 8.14.2]. It follows that

$$d_{\mathcal{C}} = \prod_{i\in I} h_i(c(1)) = \frac{(\dim \mathcal{C})^n}{\prod_{i\in I}\dim(X_i)^2},$$

where n is the cardinality of I.

Example 5.1.9 Recall from [74] that the near-group category $\mathcal{C}$ is a rigid fusion category whose simple objects except for one are invertible. Let G be the group of isomorphism classes of invertible objects in $\mathcal{C}$ and X the isomorphism class of the remaining non-invertible simple object. The Grothendieck ring $\mathrm{Gr}(\mathcal{C})$ of $\mathcal{C}$ obey the following multiplication rule:

$$g \cdot h = gh, \quad g \cdot X = X \cdot g = X, \quad X^2 = \sum_{g \in G} g + \rho X,$$

where $g, h \in G$ and ρ is a positive integer. The matrix $[c(1)]$ of left multiplication by $c(1)$ with respect to the basis $G \cup \{X\}$ of $\mathrm{Gr}(\mathcal{C})$ can be written explicitly as follows (see [85, Example 3.3]):

$$[c(1)] = \begin{bmatrix} \mathbf{M} & \mathbf{u} \\ \mathbf{u}^t & \rho^2 + 2|G| \end{bmatrix},$$

where $\mathbf{M}$ is a square matrix of size $|G|$ whose diagonal elements are all $|G| + 1$ and off-diagonal elements are all 1, $\mathbf{u}$ is a column vector of size $|G|$ whose elements are all ρ. It is easy to compute that

$$d_{\mathcal{C}} = \det[c(1)] = (4|G| + \rho^2)|G|^{|G|}.$$

Note that the Casimir number of $\mathcal{C}$ is a minimal positive integer $m_{\mathcal{C}}$ such that

$$\sum_{g \in G} gag^{-1} + XaX = m_{\mathcal{C}}, \quad \text{for some } a \in \mathrm{Gr}(\mathcal{C}).$$

Accordingly, the Casimir number $m_{\mathcal{C}}$ and the associated $a \in \mathrm{Gr}(\mathcal{C})$ can be determined separately as follows:

Case 1 If ρ is odd, then

$$a = (4|G| + \rho^2) - 2\sum_{g \in G} g - \rho X$$

and

$$m_{\mathcal{C}} = (4|G| + \rho^2)|G|.$$

Case 2 If ρ is even, then

$$a = \frac{1}{2}(4|G| + \rho^2) - \sum_{g \in G} g - \frac{\rho}{2} X$$

and

$$m_{\mathcal{C}} = \frac{1}{2}(4|G| + \rho^2)|G|.$$

We may see from this example that $m_{\mathcal{C}}$ and $d_{\mathcal{C}}$ have the same prime factors no matter what ρ is odd or even.

5.2 Applications to Verlinde modular categories

In this section, we consider the Casimir number of a Verlinde modular category $\mathcal{C}$ of rank $n+1$ introduced in [28], also see [5]. Let $\mathfrak{g}$ be a simple complex Lie algebra, n a positive integer and $q = e^{\frac{\pi i}{n+2}}$ a complex number. The *Verlinde modular category* $\mathcal{C}(\mathfrak{g}, q)$ associative with the pair $(\mathfrak{g}, q)$ is "semisimple part" of the representation category of the associated Lusztig quantum group $U_q^L(\mathfrak{g})$, more precisely, the quotient of the subcategory of tilting modules by the subcategory of negligible modules (see [28, Section 8.12.2]).

In this section, we only consider the case $\mathfrak{g} = \mathfrak{sl}_2$ and denote the Verlinde modular category $\mathcal{C}(\mathfrak{sl}_2, q)$ by $\mathcal{C}_n(q)$. The simple objects of $\mathcal{C}_n(q)$ are $X_0, X_1, \cdots, X_n$, the irreducible representations of the Lusztig quantum group $\mathfrak{u}_q(\mathfrak{sl}_2)$ (i.e., simple comodules for the quantum function algebra $O_q(SL_2)$ with highest weights $0, 1, \cdots, n$). The tensor product in $\mathcal{C}_n(q)$ is the truncation of the usual tensor product in representation category of $\mathfrak{u}_q(\mathfrak{sl}_2)$, namely, the usual tensor product $X_i \otimes X_j$ modulo a certain "negligible" part of $X_i \otimes X_j$. For instance, $\mathcal{C}_1(q) = \mathrm{Vec}^{\omega}_{\mathbb{Z}/2\mathbb{Z}}$ (where ω is the nontrivial 3-cocycle), $\mathcal{C}_2(q)$ is one of the well-known Ising model categories [25, Appendix B].

The Grothendieck ring $\mathrm{Gr}(\mathcal{C}_n(q))$ of $\mathcal{C}_n(q)$ is the truncated Verlinde ring given in [28, Example 4.10.6] whose multiplication rule is

$$X_i X_j = \sum_{l=\max\{i+j-n,0\}}^{\min\{i,j\}} X_{i+j-2l}. \tag{5.3}$$

This Grothendieck ring is a symmetric Frobenius algebra over $\mathbb{Z}$ with the bilinear form defined by $(X_i, X_j) = \delta_{i,j}$. Thus, $\{X_i, X_i \mid 0 \leqslant i \leqslant n\}$ forms a pair of dual bases of $\mathrm{Gr}(\mathcal{C}_n(q))$ with respect to the bilinear form $(-,-)$.

The Casimir operator of $\mathrm{Gr}(\mathcal{C}_n(q))$ is the map c from $\mathrm{Gr}(\mathcal{C}_n(q))$ to its center given by

$$c(x) = \sum_{i=0}^{n} X_i x X_i, \quad \text{for } x \in \mathrm{Gr}(\mathcal{C}_n(q)).$$

Since $\mathrm{Gr}(\mathcal{C}_n(q))$ is commutative, we have $c(x) = c(1)x$, where $c(1) = \sum_{i=0}^{n} X_i^2$, which is the Casimir element of $\mathrm{Gr}(\mathcal{C}_n(q))$. The Casimir number of $\mathrm{Gr}(\mathcal{C}_n(q))$ is the non-negative integer m satisfying $\mathbb{Z} \cap \mathrm{Im}c = (m)$. This number is a monoidal invariant of $\mathcal{C}_n(q)$, so is also called the Casimir number of the category $\mathcal{C}_n(q)$. By Proposition 5.1.2, we may see that the Casimir number of $\mathcal{C}_n(q)$ can be used to detect when the Grothendieck algebra $\mathrm{Gr}(\mathcal{C}_n(q)) \otimes_{\mathbb{Z}} K$ over a field K is semisimple. We want to calculate the Casimir number of $\mathcal{C}_n(q)$. Firstly, the Casimir element $c(1)$ of $\mathrm{Gr}(\mathcal{C}_n(q))$ can be described as follows:

Lemma 5.2.1 *We have* $c(1) = \sum_{j=0}^{[\frac{n}{2}]}(n+1-2j)X_{2j}$.

Proof A direct calculation shows that

$$\begin{aligned}
c(1) &= \sum_{j=0}^{[\frac{n}{2}]} X_j^2 + \sum_{j=[\frac{n}{2}]+1}^{n} X_j^2 \\
&= \sum_{j=0}^{[\frac{n}{2}]}\sum_{l=0}^{j} X_{2j-2l} + \sum_{j=[\frac{n}{2}]+1}^{n}\sum_{l=2j-n}^{j} X_{2j-2l} \quad \text{(by (5.3))} \\
&= \begin{cases} 2(X_0 + (X_0+X_2) + \cdots + (X_0+X_2+\cdots+X_{n-2})) \\ +(X_0+X_2+\cdots+X_n), & 2 \mid n \\ 2(X_0 + (X_0+X_2) + \cdots + (X_0+X_2+\cdots+X_{n-1})), & 2 \nmid n \end{cases} \\
&= \sum_{j=0}^{[\frac{n}{2}]}(n+1-2j)X_{2j}.
\end{aligned}$$

We complete the proof. □

To describe the $\mathbb{Z}$-linear expression of $c(x)$ for any $x \in \mathrm{Gr}(\mathcal{C}_n(q))$, we need some preparations. The left multiplication by X_i with respect to the basis $\{X_0, X_1, \cdots, X_n\}$ corresponds to a matrix, which is denoted by $\mathbf{X}_i$, namely,

$$X_i \begin{pmatrix} X_0 \\ X_1 \\ \vdots \\ X_n \end{pmatrix} = \mathbf{X}_i \begin{pmatrix} X_0 \\ X_1 \\ \vdots \\ X_n \end{pmatrix}.$$

We denote $\mathbf{E}_{i+1,j+1}$ for $0 \leqslant i, j \leqslant n$ the square matrix unit of order $n+1$ with $(i+1, j+1)$-entry 1, and 0 otherwise. Then the matrix $\mathbf{X}_i$ can be written explicitly as follows:

$$\begin{aligned}
\mathbf{X}_i = {} & \mathbf{E}_{1,i+1} + \mathbf{E}_{2,i+2} + \mathbf{E}_{3,i+3} + \cdots + \mathbf{E}_{n-i+1,n+1} \\
& + \mathbf{E}_{2,i} + \mathbf{E}_{3,i+1} + \mathbf{E}_{4,i+2} + \cdots + \mathbf{E}_{n-i+2,n} \\
& + \mathbf{E}_{3,i-1} + \mathbf{E}_{4,i} + \mathbf{E}_{5,i+1} + \cdots + \mathbf{E}_{n-i+3,n-1} \\
& + \cdots \\
& + \mathbf{E}_{i+1,1} + \mathbf{E}_{i+2,2} + \mathbf{E}_{i+3,3} + \cdots + \mathbf{E}_{n+1,n-i+1} \\
= {} & \sum_{s=0}^{i}\sum_{t=0}^{n-i} \mathbf{E}_{s+t+1,i+t-s+1}. \qquad (5.4)
\end{aligned}$$

Let $\delta(i)$ be the function defined over $\mathbb{Z}$ by $\delta(i)=1$ if i is even, and 0 if i is odd. Then the coefficient of X_i in the linear expression of $c(x)$ has the following form:

Proposition 5.2.2 *Let $x=\sum_{k=0}^{n}\lambda_k X_k$. Then the coefficient of X_i in the linear expression of $c(x)$ is*

$$(n+1-i)\sum_{k=0}^{i}(k+1)\delta(i+k)\lambda_k+(i+1)\sum_{k=i+1}^{n}(n-k+1)\delta(i+k)\lambda_k.$$

Proof Let $X_jX_k=\sum_{i=0}^{n}N_{jk}^{i}X_i$. Then

$$N_{jk}^{i}=(X_jX_k,X_i)=(X_iX_j,X_k)=N_{ij}^{k}$$

since the form $(-,-)$ is associative and symmetric. It follows that

$$c(x)=c(1)x=\sum_{j,k=0}^{n}(n+1-j)\delta(j)\lambda_kX_jX_k=\sum_{i,j,k=0}^{n}(n+1-j)\delta(j)\lambda_kN_{ij}^{k}X_i.$$

Thus, the coefficient of X_i in the linear expression of $c(x)$ is

$$\sum_{j,k=0}^{n}(n+1-j)\delta(j)\lambda_kN_{ij}^{k}.$$

Moreover, this coefficient can also be written as follows:

$$\begin{aligned}
&\sum_{j,k=0}^{n}(n+1-j)\delta(j)\lambda_kN_{ij}^{k}\\
&=(\lambda_0,\lambda_1,\cdots,\lambda_n)\mathbf{X}_i\begin{pmatrix}(n+1)\delta(0)\\ n\delta(1)\\ \vdots\\ \delta(n)\end{pmatrix}\\
&=\sum_{s=0}^{i}\sum_{t=0}^{n-i}(\lambda_0,\lambda_1,\cdots,\lambda_n)\mathbf{E}_{s+t+1,i+t-s+1}\begin{pmatrix}(n+1)\delta(0)\\ n\delta(1)\\ \vdots\\ \delta(n)\end{pmatrix}\quad\text{(by (5.4))}\\
&=\sum_{s=0}^{i}\sum_{t=0}^{n-i}(n+1-i-t+s)\delta(i+t-s)\lambda_{s+t}.
\end{aligned}\tag{5.5}$$

A straightforward computation shows that the coefficient of λ_k in (5.5) is

$$\lambda_k = \begin{cases} (n+1-i)(k+1)\delta(i+k), & s+t=k \leqslant i, \\ (i+1)(n-k+1)\delta(i+k), & s+t=k > i. \end{cases}$$

Thus, (5.5) is equal to

$$(n+1-i)\sum_{k=0}^{i}(k+1)\delta(i+k)\lambda_k + (i+1)\sum_{k=i+1}^{n}(n-k+1)\delta(i+k)\lambda_k.$$

This completes the proof. □

The main result of this section is presented as follows:

Theorem 5.2.3 *The Casimir number of the Verlinde modular category $\mathcal{C}_n(q)$ is* $2n+4$.

Proof Suppose $x = \sum_{k=0}^{n} \lambda_k X_k$. It follows from Proposition 5.2.2 that the coefficient α_i of X_i in the linear expression of $c(x)$ is

$$\alpha_i = (n+1-i)\sum_{k=0}^{i}(k+1)\delta(i+k)\lambda_k + (i+1)\sum_{k=i+1}^{n}(n-k+1)\delta(i+k)\lambda_k,$$

for $0 \leqslant i \leqslant n$. If $c(x) \in \mathbb{Z}$, then $\alpha_i = 0$ for $1 \leqslant i \leqslant n$. Consider the system of equations

$$\begin{cases} \alpha_n = 0, \\ \alpha_{n-2} = 0 \end{cases}$$

with variables $\lambda_0, \lambda_1, \cdots, \lambda_n$. This implies that $\lambda_n = 0$. Similarly, the system of equations

$$\begin{cases} \alpha_{n-1} = 0, \\ \alpha_{n-3} = 0 \end{cases}$$

together with $\lambda_n = 0$ induces that $\lambda_{n-1} = 0$. Repeating this argument until the following equations:

$$\begin{cases} \alpha_3 = 0, \\ \alpha_1 = 0, \end{cases}$$

we obtain that $\lambda_n = \lambda_{n-1} = \cdots = \lambda_3 = 0$. Now consider the system of equations

$$\begin{cases} \alpha_2 = 0, \\ \alpha_1 = 0. \end{cases}$$

It follows that $\lambda_1 = 0$ and $\lambda_0 = -3\lambda_2$. Thus, the coefficient α_0 of X_0 is

$$\begin{aligned}\alpha_0 &= (n+1)\lambda_0 + \sum_{k=1}^{n}(n-k+1)\delta(k)\lambda_k \\ &= (n+1)\lambda_0 + (n-1)\lambda_2 \\ &= -(2n+4)\lambda_2.\end{aligned}$$

We conclude that the Casimir number of $\mathcal{C}_n(q)$ is $2n+4$. □

Remark 5.2.4 *The maximal non-negative eigenvalue of the matrix* $\mathbf{X}_i$ *is called the Frobenius-Perron dimension of* X_i*, denoted by* $\mathrm{FPdim}(X_i)$*. Then* FPdim *induces an algebra morphism from* $\mathrm{Gr}(\mathcal{C}_n(q))$ *to the field* $\mathbb{C}$ *of complex numbers. Note that*

$$\mathrm{FPdim}(X_i) = \frac{q^{i+1} - q^{-i-1}}{q - q^{-1}},$$

for $0 \leqslant i \leqslant n$ *(see [28, Exercise 4.10.7]). The Frobenius-Perron dimension of* $\mathcal{C}_n(q)$ *is the number* $\mathrm{FPdim}(\mathcal{C}_n(q))$ *defined by*

$$\mathrm{FPdim}(\mathcal{C}_n(q)) := \mathrm{FPdim}(c(1)) = \sum_{i=0}^{n}(\mathrm{FPdim}(X_i))^2 = \frac{2n+4}{(q-q^{-1})^2}.$$

This shows that

$$2n+4 = \mathrm{FPdim}(\mathcal{C}_n(q))(q-q^{-1})^2,$$

which gives a relationship between the Casimir number of $\mathcal{C}_n(q)$ *and the Frobenius-Perron dimension of* $\mathcal{C}_n(q)$*. By the way, this equality can be obtained as well from the following approach: it follows from Theorem* 5.2.3 *that* $2n+4 = c(1)x$*, where* $x = 3 - X_2$*, applying* FPdim *to this equality we also obtain that*

$$2n+4 = \mathrm{FPdim}(\mathcal{C}_n(q))\mathrm{FPdim}(x) = \mathrm{FPdim}(\mathcal{C}_n(q))(q-q^{-1})^2.$$

The Casimir number of $\mathcal{C}_n(q)$ *can be used to determine the semisimplicity of the Grothendieck algebra* $\mathrm{Gr}(\mathcal{C}_n(q)) \otimes_{\mathbb{Z}} K$ *over a field* K*. Next, we turn to consider the Jacobson radical of* $\mathrm{Gr}(\mathcal{C}_n(q)) \otimes_{\mathbb{Z}} K$ *in the case when* $2n+4$ *is zero in* K*. To this end, we need to present the Grothendieck ring* $\mathrm{Gr}(\mathcal{C}_n(q))$ *in terms of generators and relations.*

Let $\mathbb{Z}[X]$ *be a polynomial ring with one variable* X *over* $\mathbb{Z}$ *and* $(E_{n+1}(X))$ *the ideal of* $\mathbb{Z}[X]$ *generated by the* $n+1$*-th Dickson polynomial* $E_{n+1}(X)$ *(see* (4.1)*). The image of a polynomial* $f(X)$ *under the canonical ring epimorphism* $\mathbb{Z}[X] \to \mathbb{Z}[X]/(E_{n+1}(X))$ *is denoted by* $\overline{f(X)}$*. We have the following lemma:*

Lemma 5.2.5 *For* $0 \leqslant i, j \leqslant n$, *the equality*

$$\overline{E_i(X)E_j(X)} = \sum_{l=\max\{i+j-n,0\}}^{\min\{i,j\}} \overline{E_{i+j-2l}(X)}$$

holds in $\mathbb{Z}[X]/(E_{n+1}(X))$.

Proof We suppose that $E_s(X) = 0$ if $s < 0$. We proceed by induction on $i + j$ only for the case $0 \leqslant i + j \leqslant n$, and the proof of the case $n \leqslant i + j \leqslant 2n$ is similar. It is obvious that the identity holds for $i + j = 0$. For a fixed $1 \leqslant k \leqslant n - 1$, suppose that the identity holds for $1 \leqslant i + j \leqslant k$. We show that it also holds for the case $i + j = k + 1$. Note that $(i - 1) + j \leqslant k$ and $(i - 2) + j \leqslant k$. Applying the induction hypothesis on $(i - 1) + j \leqslant k$ and $(i - 2) + j \leqslant k$, we obtain the following two equalities:

$$\overline{E_{i-1}(X)E_j(X)} = \sum_{l=\max\{i-1+j-n,0\}}^{\min\{i-1,j\}} \overline{E_{i-1+j-2l}(X)} \tag{5.6}$$

and

$$\overline{E_{i-2}(X)E_j(X)} = \sum_{l=\max\{i-2+j-n,0\}}^{\min\{i-2,j\}} \overline{E_{i-2+j-2l}(X)}. \tag{5.7}$$

Now consider the product $\overline{XE_{i-1}(X)E_j(X)}$ in $\mathbb{Z}[X]/(E_{n+1}(X))$. On the one hand, using (5.6) we get that

$$\begin{aligned}\overline{XE_{i-1}(X)E_j(X)} &= \overline{X} \sum_{l=\max\{i-1+j-n,0\}}^{\min\{i-1,j\}} \overline{E_{i-1+j-2l}(X)} \\ &= \sum_{l=\max\{i-1+j-n,0\}}^{\min\{i-1,j\}} (\overline{E_{i+j-2l}(X)} + \overline{E_{i-2+j-2l}(X)}).\end{aligned}$$

On the other hand, using (5.7) we have

$$\begin{aligned}\overline{XE_{i-1}(X)E_j(X)} &= (\overline{E_i(X)} + \overline{E_{i-2}(X)})\overline{E_j(X)} \\ &= \overline{E_i(X)E_j(X)} + \sum_{l=\max\{i-2+j-n,0\}}^{\min\{i-2,j\}} \overline{E_{i-2+j-2l}(X)}.\end{aligned}$$

It follows that

$$\overline{E_i(X)E_j(X)} = \sum_{l=\max\{i-1+j-n,0\}}^{\min\{i-1,j\}} (\overline{E_{i+j-2l}(X)} + \overline{E_{i-2+j-2l}(X)})$$

$$-\sum_{l=\max\{i-2+j-n,0\}}^{\min\{i-2,j\}} \overline{E_{i-2+j-2l}(X)}.$$

By discussing the cases $i-1<j$, $i-1=j$ and $i-1>j$ separately, we obtain that

$$\overline{E_i(X)E_j(X)} = \sum_{l=\max\{i+j-n,0\}}^{\min\{i,j\}} \overline{E_{i+j-2l}(X)}.$$

We complete the proof. □

Theorem 5.2.6 *The Grothendieck ring* $\mathrm{Gr}(\mathcal{C}_n(q))$ *is isomorphic to the quotient ring* $\mathbb{Z}[X]/(E_{n+1}(X))$.

Proof Consider the $\mathbb{Z}$-linear map θ from $\mathrm{Gr}(\mathcal{C}_n(q))$ to $\mathbb{Z}[X]/(E_{n+1}(X))$ given by $\theta(X_i)=\overline{E_i(X)}$ for $0\leqslant i\leqslant n$. This is a ring epimorphism by Lemma 5.2.5. To see that this map is injective, we suppose $\sum_{i=0}^n \lambda_i\overline{E_i(X)}=0$ for each $\lambda_i\in\mathbb{Z}$, then

$$\sum_{i=0}^n \lambda_i E_i(X) = E_{n+1}(X)f(X), \quad \text{for some } f(X)\in\mathbb{Z}[X].$$

By comparing the degrees of both two sides of the equality, we obtain that $f(X)=0$, and hence $\lambda_i=0$ for $0\leqslant i\leqslant n$. □

The factorizations of $E_{n+1}(X)$ have been carried out using much lengthier methods by Chou [17] (see also [9]). According to Theorem 5.2.6, we obtain the following criterion for $E_{n+1}(X)$ without multiple factors in $K[X]$:

Proposition 5.2.7 *The* $(n+1)$*-th Dickson polynomial* $E_{n+1}(X)$ *of the second kind has no multiple factors in* $K[X]$ *if and only if* $2n+4$ *is a unit in* K.

Proof We have by Theorem 5.2.6 that

$$\mathrm{Gr}(\mathcal{C}_n(q))\otimes_{\mathbb{Z}} K \cong K[X]/(E_{n+1}(X)).$$

It follows from Theorem 5.2.3 that $2n+4$ is a unit in K if and only if $K[X]/(E_{n+1}(X))$ is semisimple, if and only if $E_{n+1}(X)$ has no multiple factors in $K[X]$, as desired. □

We turn to the study of the Jacobson radical of $\mathrm{Gr}(\mathcal{C}_n(q))\otimes_{\mathbb{Z}} K$, or equivalently, $K[X]/(E_{n+1}(X))$ in the case $2n+4$ is zero in K. Note that a product of all distinct irreducible factors of $E_{n+1}(X)$ gives rise to a generator of the Jacobson radical of $K[X]/(E_{n+1}(X))$.

Proposition 5.2.8 *Suppose that the characteristic of* K *is* $p>2$. *If* $p\mid 2n+4$, *write* $n+2=p^r(m+1)$ *where* $(p,m+1)=1$, *then the Jacobson radical of* $K[X]/(E_{n+1}(X))$ *is a principal ideal generated by* $\overline{E_m(X)(X^2-4)}$.

Proof The decomposition

$$E_{n+1}(X)=E_m(X)^{p^r}(X^2-4)^{\frac{p^r-1}{2}}$$

holds in $K[X]$ (see [9, Section 3]), and the Dickson polynomial $E_m(X)$ has no multiple factors in $K[X]$ by Proposition 5.2.7. It follows that $E_m(X)(X^2-4)$ is a product of all distinct irreducible factors of $E_{n+1}(X)$. We conclude that the Jacobson radical of $K[X]/(E_{n+1}(X))$ is a principal ideal generated by $\overline{E_m(X)(X^2-4)}$. □

Suppose that the characteristic of K is 2. Then the factorizations of Dickson polynomials of the second kind are a little bit more complicated. If m is even, it follows from [9, Theorem 6] that $E_m(X) = F_m(X)^2$, where

$$F_m(X) = \sum_{j=0}^{\frac{m}{2}} \binom{m-j}{j} (-1)^j X^{\frac{m}{2}-j},$$

which is a product of several distinct irreducible polynomials in $K[X]$, which occur in cliques corresponding to the divisors d of $m+1$ with $d > 1$. For each such d there correspond $\varphi(d)/(2k_d)$ irreducible factors, where φ is the Euler's totient function and k_d is the least positive integer such that $q^{k_d} \equiv \pm 1 \pmod{d}$. Each of such irreducible factors has the form

$$\prod_{i=0}^{k_d-1} (X - (\zeta_d^{2^i} + \zeta_d^{-2^i})),$$

for some choice of ζ_d, where ζ_d is a primitive d-th root of unity.

Proposition 5.2.9 *Suppose the characteristic of K is 2.*

(1) *If $n+1$ is even, then the Jacobson radical of $K[X]/(E_{n+1}(X))$ is a principal ideal generated by $\overline{F_{n+1}(X)}$.*

(2) *If $n+1$ is odd, write $n+2 = 2^r(m+1)$, where m is even, then the Jacobson radical of $K[X]/(E_{n+1}(X))$ is a principal ideal generated by $\overline{XF_m(X)}$.*

Proof (1) If $n+1$ is even, then $F_{n+1}(X)$ is a product of all distinct irreducible factors of $E_{n+1}(X)$ as stated above. Thus, the Jacobson radical of $K[X]/(E_{n+1}(X))$ is a principal ideal generated by $\overline{F_{n+1}(X)}$.

(2) If $n+1$ is odd, write $n+2 = 2^r(m+1)$, where $r \geqslant 1$ and m is even. In this case, by [9, Section 3] we have

$$E_{n+1}(X) = X^{2^r-1} E_m(X)^{2^r} = X^{2^r-1} F_m(X)^{2^{r+1}}.$$

Thus, $XF_m(X)$ can be written as a product of all distinct irreducible factors of $E_{n+1}(X)$, and hence the Jacobson radical of $K[X]/(E_{n+1}(X))$ is a principal ideal generated by $\overline{XF_m(X)}$. □

5.3 Prime factors of Casimir numbers

The representation category H-mod of a finite dimensional semisimple Hopf algebra H is a fusion category. Denote by m_H and d_H respectively the Casimir number and the determinant of the fusion category H-mod. We also denote by $m_{\mathcal{C}}$ the Casimir number of a fusion category $\mathcal{C}$. In this section, we will give some results concerning prime factors of m_H and $m_{\mathcal{C}}$.

A finite dimensional Hopf algebra H is call *pivotal* if H contains a group-like element g such that $S^2(h) = ghg^{-1}$ for $h \in H$. The representation category H-mod of a finite dimensional semisimple pivotal Hopf algebra H is a pivotal fusion category.

Proposition 5.3.1 *Let H be a finite dimensional semisimple pivotal Hopf algebra over $\Bbbk$. The Casimir number $m_H \neq 0$ in $\Bbbk$ if and only if $S^2 = id$ and* $\dim_{\Bbbk}(H) \neq 0$ *in* $\Bbbk$.

Proof The Casimir number m_H is not zero in $\Bbbk$ if and only if H-mod is non-degenerate by Proposition 5.1.3, if and only if H is cosemisimple by [30, Section 9.1], if and only if $S^2 = id$ and $\dim_{\Bbbk}(H) \neq 0$ in $\Bbbk$ by [27, Corollary 3.2]. □

Since a finite dimensional semisimple and cosemisimple Hopf algebra H always satisfies that $S^2 = id$ and $\dim_{\Bbbk}(H) \neq 0$ in $\Bbbk$ (see [27, Corollary 3.2]), Proposition 5.3.1 has the following corollary:

Corollary 5.3.2 *Let H be a finite dimensional semisimple and cosemisimple Hopf algebra over a field $\Bbbk$. The Casimir number m_H is not zero in $\Bbbk$.*

The following result gives more information about the Casimir number m_H of a semisimple and cosemisimple Hopf algebra H under a certain hypothesis.

Proposition 5.3.3 *Let H be a finite dimensional semisimple and cosemisimple Hopf algebra over a field $\Bbbk$. If the Grothendieck ring $G_0(H)$ of H is commutative, then the Casimir number m_H, the determinant d_H and the dimension* $\dim_{\Bbbk}(H)$ *have the same prime factors.*

Proof According to Corollary 5.1.6, we know that m_H and d_H have the same prime factors. We only need to show that d_H and $\dim_{\Bbbk}(H)$ have the same prime factors.

We first consider the case $\mathrm{char}(\Bbbk) = 0$. The set of isomorphism classes of simple objects of H-mod is denoted by $\{X_i\}_{i\in I}$. Since the Grothendieck ring $G_0(H)$ is commutative, it follows from [52, Proposition 20] that all eigenvalues of the matrix $[c(1)]$ of left multiplication by $c(1) = \sum_{i\in I} X_i X_{i^*}$ with respect to the basis $\{X_i\}_{i\in I}$ of $G_0(H)$ are positive integers, and moreover, all these eigenvalues divide $\dim_{\Bbbk}(H)$. In particular, $\dim_{\Bbbk}(H)$ itself is the largest eigenvalue of $[c(1)]$ (see [52, Proposition 8]). On the other hand, the determinant d_H is obtained by multiplying all these eigenvalues. Thus, d_H and $\dim_{\Bbbk}(H)$ have the same prime factors.

For the case $\mathrm{char}(\Bbbk) = p > 0$, we denote by $\mathcal{O}$ the ring of Witt vectors of $\Bbbk$ and K the field of fractions of $\mathcal{O}$. For the semisimple and cosemisimple Hopf algebra H, using the lifting Theorem [27, Theorem 2.1] we may construct a Hopf algebra A over $\mathcal{O}$ which is free of rank $\dim_{\Bbbk}(H)$ as an $\mathcal{O}$-module such that A/pA is isomorphic to H

as a Hopf algebra. The Hopf algebra $A_0 := A \otimes_{\mathcal{O}} K$ is a semisimple and cosemisimple Hopf algebra over the field K of characteristic 0 with the same Grothendieck ring as for H. It follows that the Grothendieck ring $G_0(A_0)$ is commutative and the determinant d_{A_0} of A_0-mod is equal to the determinant d_H of H-mod. By the same argument as for the case of $\mathrm{char}(\Bbbk) = 0$, we may see that the determinant d_{A_0} and $\dim_K(A_0)$ have the same prime factors. Note that $\dim_K(A_0) = \dim_K(A \otimes_{\mathcal{O}} K)$ which is equal to $\dim_{\Bbbk}(H)$ since the Hopf algebra A over $\mathcal{O}$ is free of rank $\dim_{\Bbbk}(H)$ and $\mathcal{O}$ as a discrete valuation ring is a unique factorization domain. We conclude that d_H and $\dim_{\Bbbk}(H)$ have the same prime factors. □

Applying Proposition 5.3.3 to the Drinfeld double of a semisimple and cosemisimple Hopf algebra, we have the following result:

Theorem 5.3.4 *Let H be a finite dimensional semisimple and cosemisimple Hopf algebra over a field $\Bbbk$ and $D(H)$ the Drinfeld double of H. The Casimir number $m_{D(H)}$, the determinant $d_{D(H)}$ and the dimension $\dim_{\Bbbk}(H)$ have the same prime factors.*

Proof The representation category of the Drinfeld double $D(H)$ is a modular fusion category over $\Bbbk$, since $D(H)$ is a quasitriangular semisimple and cosemisimple Hopf algebra (see [56, Corollary 10.3.13]). It follows that the Grothendieck ring $G_0(D(H))$ of $D(H)$ is a commutative ring. By Proposition 5.3.3, the Casimir number $m_{D(H)}$, the determinant $d_{D(H)}$ and the dimension $\dim_{\Bbbk} D(H) = \dim_{\Bbbk}(H)^2$ have the same prime factors. This gives the desired result. □

Next, we describe a relationship between the Casimir number $m_{\mathcal{C}}$ (resp. the determinant $d_{\mathcal{C}}$) of a fusion category $\mathcal{C}$ and the Casimir number $m_{\widetilde{\mathcal{C}}}$ (resp. the determinant $d_{\widetilde{\mathcal{C}}}$) of $\widetilde{\mathcal{C}}$, where $\widetilde{\mathcal{C}}$ is the pivotalization of $\mathcal{C}$ stated below.

Let $\mathcal{C}$ be a fusion category over a field $\Bbbk$. Recall from [30, Theorem 2.6] that there exists an isomorphism $\gamma : id \to (-)^{****}$ between the identity and the fourth duality tensor autoequivalences of $\mathcal{C}$. Denote by $\widetilde{\mathcal{C}} := \mathcal{C}^{\mathbb{Z}/2\mathbb{Z}}$ the corresponding equivariantization. More explicitly, simple objects of $\widetilde{\mathcal{C}}$ are pairs (X, α), where X is a simple object of $\mathcal{C}$, and $\alpha : X \to X^{**}$ satisfies $\alpha^{**}\alpha = \gamma_X$. The fusion category $\widetilde{\mathcal{C}}$ has a canonical pivotal structure which is called the *pivotalization* of $\mathcal{C}$ (see [28, Definition 7.21.9] for details). Moreover, the pivotal fusion category $\widetilde{\mathcal{C}}$ is also spherical (see [29, Corollary 7.6]).

To describe any simple object (X, α) of $\widetilde{\mathcal{C}}$, we first fix an isomorphism $\theta : X \to X^{**}$. Since $\mathrm{Hom}(X, X^{**})$ is one dimensional, we may write $\alpha = u\theta$ and $\gamma_X = v\theta^{**}\theta$ for some $u, v \in \Bbbk^{\times}$. Then $\alpha^{**}\alpha = \gamma_X$ implies that $u^2 = v$. Therefore, for each simple object X of $\mathcal{C}$, we only have two choices of α, and if one of them is α then another one is $-\alpha$. In view of this, we may write $(X, \alpha) = X^+$ and $(X, -\alpha) = X^-$. It follows that $\mathbf{1}^+ = \mathbf{1}$, $\mathbf{1}^- \otimes \mathbf{1}^- = \mathbf{1}$, $\dim(\mathbf{1}^-) = -1$, and $X^{\pm} \otimes \mathbf{1}^- = \mathbf{1}^- \otimes X^{\pm} = X^{\mp}$ (see [67, Section 5.1]). Note that the forgetful function $F : \widetilde{\mathcal{C}} \to \mathcal{C}, X^{\pm} \mapsto X$

preserves squared norms of simple objects [28, Remark 7.21.11]. It follows that $\dim(\widetilde{\mathcal{C}}) = 2\dim(\mathcal{C})$.

If $\mathrm{char}(\Bbbk) \neq 2$, the Grothendieck algebra $\mathrm{Gr}(\widetilde{\mathcal{C}}) \otimes_{\mathbb{Z}} \Bbbk$ has the following decomposition:

$$\mathrm{Gr}(\widetilde{\mathcal{C}}) \otimes_{\mathbb{Z}} \Bbbk = e(\mathrm{Gr}(\widetilde{\mathcal{C}}) \otimes_{\mathbb{Z}} \Bbbk) \oplus (1-e)(\mathrm{Gr}(\widetilde{\mathcal{C}}) \otimes_{\mathbb{Z}} \Bbbk), \tag{5.8}$$

where $e = \dfrac{\mathbf{1} - \mathbf{1}^-}{2}$, which is a central idempotent element of $\mathrm{Gr}(\widetilde{\mathcal{C}}) \otimes_{\mathbb{Z}} \Bbbk$. It follows from [67, Section 5.1] that

$$\mathrm{Gr}(\mathcal{C}) \otimes_{\mathbb{Z}} \Bbbk \cong (\mathrm{Gr}(\widetilde{\mathcal{C}}) \otimes_{\mathbb{Z}} \Bbbk)/e(\mathrm{Gr}(\widetilde{\mathcal{C}}) \otimes_{\mathbb{Z}} \Bbbk) \cong (1-e)(\mathrm{Gr}(\widetilde{\mathcal{C}}) \otimes_{\mathbb{Z}} \Bbbk).$$

The numbers $m_{\mathcal{C}}$ (resp. $d_{\mathcal{C}}$) and $m_{\widetilde{\mathcal{C}}}$ (resp. $d_{\widetilde{\mathcal{C}}}$) have the following divisibility relation:

Proposition 5.3.5 *Let $\mathcal{C}$ be a fusion category over a field $\Bbbk$ and $\widetilde{\mathcal{C}}$ the pivotalization of $\mathcal{C}$.*

(1) $m_{\mathcal{C}} \mid m_{\widetilde{\mathcal{C}}}$.

(2) $d_{\mathcal{C}} \mid d_{\widetilde{\mathcal{C}}}$.

Proof (1) Denote by $\{X_i\}_{i\in I}$ the set of isomorphism classes of simple objects of $\mathcal{C}$. Then $\{X_i^{\pm}\}_{i\in I}$ is the set of isomorphism classes of simple objects of $\widetilde{\mathcal{C}}$. For the Casimir number $m_{\widetilde{\mathcal{C}}}$, there exists some $a \in \mathrm{Gr}(\widetilde{\mathcal{C}})$ such that

$$\sum_{i\in I} X_i^{\pm} a (X_i^{\pm})^* = m_{\widetilde{\mathcal{C}}}.$$

Applying the ring homomorphism $f : \mathrm{Gr}(\widetilde{\mathcal{C}}) \to \mathrm{Gr}(\mathcal{C})$ induced by the forgetful function $F : \widetilde{\mathcal{C}} \to \mathcal{C}$ to this equation, we have

$$\sum_{i\in I} X_i 2f(a)(X_i)^* = m_{\widetilde{\mathcal{C}}}.$$

It follows that

$$m_{\widetilde{\mathcal{C}}} \in \mathbb{Z} \cap \mathrm{Im}c = (m_{\mathcal{C}}).$$

This gives a proof of Part (1).

(2) In the Grothendieck ring $\mathrm{Gr}(\widetilde{\mathcal{C}})$ we suppose for any $j \in I$ that

$$\sum_{i\in I} X_i^+ (X_i^+)^* X_j^+ = \sum_{i\in I} \mu_{ij} X_i^+ + \sum_{i\in I} \nu_{ij} X_i^-, \tag{5.9}$$

where $\mu_{ij}, \nu_{ij} \in \mathbb{Z}$. Then for any $j \in I$, we have

$$\sum_{i\in I} X_i^+ (X_i^+)^* X_j^- = \sum_{i\in I} X_i^+ (X_i^+)^* X_j^+ \mathbf{1}^- = \sum_{i\in I} \mu_{ij} X_i^- + \sum_{i\in I} \nu_{ij} X_i^+.$$

This means that, in the Grothendieck ring $\mathrm{Gr}(\widetilde{\mathcal{C}})$, the matrix of left multiplication by the Casimir element

$$\sum_{i\in I} X_i^{\pm}(X_i^{\pm})^* = 2\sum_{i\in I} X_i^{+}(X_i^{+})^*$$

with respect to the basis $\{X_i^{\pm}\}_{i\in I}$ of $\mathrm{Gr}(\widetilde{\mathcal{C}})$ is

$$2\begin{pmatrix} \mathbf{A} & \mathbf{B} \\ \mathbf{B} & \mathbf{A} \end{pmatrix},$$

where $\mathbf{A} = (\mu_{ij})_{n\times n}$, $\mathbf{B} = (\nu_{ij})_{n\times n}$ and n is the cardinality of I. Thus, the determinant of $\widetilde{\mathcal{C}}$ is

$$d_{\widetilde{\mathcal{C}}} = 2^{2n}\det\begin{pmatrix} \mathbf{A} & \mathbf{B} \\ \mathbf{B} & \mathbf{A} \end{pmatrix} = 2^{2n}\det(\mathbf{A}+\mathbf{B})\det(\mathbf{A}-\mathbf{B}).$$

Applying the ring homomorphism $f : \mathrm{Gr}(\widetilde{\mathcal{C}}) \to \mathrm{Gr}(\mathcal{C})$ induced from the forgetful function $F : \widetilde{\mathcal{C}} \to \mathcal{C}$ to the equation (5.9), we have that

$$\sum_{i\in I} X_i(X_i)^* X_j = \sum_{i\in I} (\mu_{ij} + \nu_{ij}) X_i.$$

This shows that, in the Grothendieck ring $\mathrm{Gr}(\mathcal{C})$, the matrix of left multiplication by the Casimir element $\sum_{i\in I} X_i X_i^*$ with respect to the basis $\{X_i\}_{i\in I}$ of $\mathrm{Gr}(\mathcal{C})$ is $\mathbf{A}+\mathbf{B}$. Thus, the determinant of $\mathcal{C}$ is $d_{\mathcal{C}} = \det(\mathbf{A}+\mathbf{B})$, which is a factor of $d_{\widetilde{\mathcal{C}}}$. □

As a consequence, we have the following result:

Proposition 5.3.6 *Let $\mathcal{C}$ be a fusion category over a field $\Bbbk$ with* $\mathrm{char}(\Bbbk) \neq 2$. *If $\mathcal{C}$ is non-degenerate, then $m_{\mathcal{C}} \neq 0$ in $\Bbbk$.*

Proof Since $\mathcal{C}$ is non-degenerate, i.e., the global dimension $\dim(\mathcal{C}) \neq 0$ in $\Bbbk$, it follows that $\dim(\widetilde{\mathcal{C}}) = 2\dim(\mathcal{C}) \neq 0$. Thus, the pivotal fusion category $\widetilde{\mathcal{C}}$ is non-degenerate. It follows from Proposition 5.1.3 that $m_{\widetilde{\mathcal{C}}} \neq 0$ in $\Bbbk$. As a result, $m_{\mathcal{C}} \neq 0$ since $m_{\mathcal{C}}$ is a factor of $m_{\widetilde{\mathcal{C}}}$. □

We expect that the converse of Proposition 5.3.6 is also true. However, the proof seems too hard to be finished. What we can do is the proof of the following statement:

Proposition 5.3.7 *Let $\mathcal{C}$ be a fusion category over a field $\Bbbk$ with* $\mathrm{char}(\Bbbk) \neq 2$. *Then $\mathcal{C}$ is non-degenerate if and only if the subalgebra $e(\mathrm{Gr}(\widetilde{\mathcal{C}}) \otimes_{\mathbb{Z}} \Bbbk)$ is semisimple, where $e = \dfrac{\mathbf{1} - \mathbf{1}^-}{2}$ is a central idempotent element of* $\mathrm{Gr}(\widetilde{\mathcal{C}}) \otimes_{\mathbb{Z}} \Bbbk$.

Proof The global dimension $\dim(\mathcal{C}) \neq 0$ shows that $\dim(\widetilde{\mathcal{C}}) = 2\dim(\mathcal{C}) \neq 0$. Thus, the Grothendieck algebra $\mathrm{Gr}(\widetilde{\mathcal{C}}) \otimes_{\mathbb{Z}} \Bbbk$ of the pivotal fusion category $\widetilde{\mathcal{C}}$ is semisimple by [73, Theorem 6.5]. It follows that the quotient algebra (see (5.8))

$$(\mathrm{Gr}(\widetilde{\mathcal{C}}) \otimes_{\mathbb{Z}} \Bbbk)/(1-e)(\mathrm{Gr}(\widetilde{\mathcal{C}}) \otimes_{\mathbb{Z}} \Bbbk) \cong e(\mathrm{Gr}(\widetilde{\mathcal{C}}) \otimes_{\mathbb{Z}} \Bbbk)$$

is semisimple. Conversely, consider

$$t = \sum_{i\in I} \dim(X_i^{\pm})(X_i^{\pm})^* \in \mathrm{Gr}(\widetilde{\mathcal{C}}) \otimes_{\mathbb{Z}} \Bbbk.$$

Obviously, $t \neq 0$. For any $a \in \mathrm{Gr}(\widetilde{\mathcal{C}}) \otimes_{\mathbb{Z}} \Bbbk$, it follows from (5.1) that

$$ta = \sum_{i\in I} \dim(X_i^{\pm})(X_i^{\pm})^* a = \sum_{i\in I} \dim(aX_i^{\pm})(X_i^{\pm})^* = \dim(a)t.$$

Similarly, it follows from (5.2) that $at = \dim(a)t$. Thus, t is an central element of $\mathrm{Gr}(\widetilde{\mathcal{C}}) \otimes_{\mathbb{Z}} \Bbbk$ satisfying

$$t^2 = \dim(t)t = \dim(\widetilde{\mathcal{C}})t = 2\dim(\mathcal{C})t. \tag{5.10}$$

Moreover,

$$t = et + (1-e)t = et + \dim(1-e)t = et \in e(\mathrm{Gr}(\widetilde{\mathcal{C}}) \otimes_{\mathbb{Z}} \Bbbk).$$

If $\dim(\mathcal{C}) = 0$, then $t^2 = 0$ by (5.10) and hence the ideal of $e(\mathrm{Gr}(\widetilde{\mathcal{C}}) \otimes_{\mathbb{Z}} \Bbbk)$ generated by t is nilpotent, a contradiction to the semisimplicity of $e(\mathrm{Gr}(\widetilde{\mathcal{C}}) \otimes_{\mathbb{Z}} \Bbbk)$. The proof is completed. □

5.4 Casimir numbers vs. Frobenius-Schur exponents

In this section, we shall show that the Casimir number and the Frobenius-Schur exponent of a spherical fusion category over the field $\mathbb{C}$ of complex numbers have the same prime factors.

Let $\mathcal{C}$ be a fusion category over a field $\Bbbk$ and V a finite dimensional left $\mathrm{Gr}(\mathcal{C}) \otimes_{\mathbb{Z}} K$-module over an algebraic closure field K. For any $\varphi \in \mathrm{End}_K(V)$, we define $\mathcal{I}(\varphi) \in \mathrm{End}_K(V)$ by

$$\mathcal{I}(\varphi)(v) = \sum_{i\in I} X_i \varphi(X_{i^*} v), \quad \text{for } v \in V.$$

Then $\mathcal{I}(\varphi)$ lies in $\mathrm{End}_{\mathrm{Gr}(\mathcal{C})\otimes_{\mathbb{Z}} K}(V)$ and does not depend on the choice of a pair of dual bases of $\mathrm{Gr}(\mathcal{C}) \otimes_{\mathbb{Z}} K$ (see [32, Lemma 7.1.10]). If V is a simple $\mathrm{Gr}(\mathcal{C}) \otimes_{\mathbb{Z}} K$-module, then $\mathrm{End}_{\mathrm{Gr}(\mathcal{C})\otimes_{\mathbb{Z}} K}(V) \cong K$. In this case, there exists a unique element $c_V \in K$ such that

$$\mathcal{I}(\varphi) = c_V \mathrm{Tr}(\varphi) id_V, \quad \text{for all } \varphi \in \mathrm{End}_K(V).$$

Such an element c_V only depends on the isomorphism class of V and is called the *Schur element* associated with V (see [32, Theorem 7.2.1]). Note that the semisimplicity criterion stated in [32, Theorem 7.2.6] works for Grothendieck algebras. Namely, the Grothendieck algebra $\mathrm{Gr}(\mathcal{C}) \otimes_{\mathbb{Z}} K$ is semisimple if and only if any Schur element associated with a simple module over $\mathrm{Gr}(\mathcal{C}) \otimes_{\mathbb{Z}} K$ is not zero in K.

Let V be a simple $\mathrm{Gr}(\mathcal{C}) \otimes_{\mathbb{Z}} K$-module with the Schur element c_V. The character of V is denoted by χ_V. Then $\sum_{i\in I} \chi_V(X_i)X_{i^*}$ is a central element of $\mathrm{Gr}(\mathcal{C}) \otimes_{\mathbb{Z}} K$. This element acts by a scalar f_V on V and by zero on any simple module not isomorphic to V. The scalar f_V is called the *formal codegree* of V (see [66, Lemma 2.3]).

Lemma 5.4.1 *Let V be a simple $\mathrm{Gr}(\mathcal{C}) \otimes_{\mathbb{Z}} K$-module with the Schur element c_V and the formal codegree f_V. The action of $\sum_{i,j\in I} X_iX_jX_{i^*}X_{j^*}$ on V is a scalar multiple by $c_V f_V$.*

Proof For any simple module V over $\mathrm{Gr}(\mathcal{C}) \otimes_{\mathbb{Z}} K$, there is a corresponding algebra morphism

$$\rho_V : \mathrm{Gr}(\mathcal{C}) \otimes_{\mathbb{Z}} K \to \mathrm{End}_K(V), \quad \rho_V(a)(v) = av, \quad \text{for } a \in \mathrm{Gr}(\mathcal{C}) \otimes_{\mathbb{Z}} K,\ v \in V.$$

Note that $\mathcal{I}(\varphi)(v) = c_V \mathrm{Tr}(\varphi)v$ holds for any $\varphi \in \mathrm{End}_K(V)$ and $v \in V$. Replacing φ and v in this equality by $\rho_V(X_j)$ and $X_{j^*}v$ respectively, we have

$$\mathcal{I}(\rho_V(X_j))(X_{j^*}v) = c_V \mathrm{Tr}(\rho_V(X_j))X_{j^*}v.$$

Summing over all $j \in I$ we have

$$\sum_{j\in I} \mathcal{I}(\rho_V(X_j))(X_{j^*}v) = c_V \sum_{j\in I} \mathrm{Tr}(\rho_V(X_j))X_{j^*}v.$$

Taking into account the definition of $\mathcal{I}$, we have

$$\sum_{i,j\in I} X_i\rho_V(X_j)(X_{i^*}X_{j^*}v) = c_V \sum_{j\in I} \chi_V(X_j)X_{j^*}v.$$

This gives rise to the desired result

$$\sum_{i,j\in I} X_iX_jX_{i^*}X_{j^*}v = c_V f_V v, \quad \text{for any } v \in V.$$ □

Note that $\mathrm{Gr}(\mathcal{C}) \otimes_{\mathbb{Z}} \mathbb{C}$ is always semisimple and $\dim(\mathcal{C})$ is always not zero in the field $\mathbb{C}$ of complex numbers. Thus, [73, Theorem 6.5] is trivial if the ground field $\Bbbk$ is taken to be $\mathbb{C}$. Next, we shall give a modification version of [73, Theorem 6.5] so that we can use it to present another statement of the Cauchy theorem for spherical fusion categories.

Let $\mathcal{C}$ be a spherical fusion category over $\mathbb{C}$ with isomorphism classes of simple objects $\{X_i\}_{i\in I}$. The Frobenius-Schur exponent of $\mathcal{C}$ has been defined in [61, Definition 5.1] in terms of the higher Frobenius-Schur indicators of objects of $\mathcal{C}$. This exponent, denoted by N, can be regarded as the order of the twist θ of the Drinfeld center $Z(\mathcal{C})$ associated with a pivotal structure of $\mathcal{C}$ (see [61, Theorem 5.5]). Let $\xi_N \in \mathbb{C}$ be a primitive N-th root of unity. Then $\mathbb{Z}[\xi_N]$ is a Dedekind domain and every nonzero proper ideal factors into a product of prime ideal factors. Let $\mathfrak{p}$ be a prime ideal of $\mathbb{Z}[\xi_N]$. Then $\mathfrak{p}$ is maximal since $\mathbb{Z}[\xi_N]$ is Dedekind. Thus, the quotient ring $\mathbb{Z}[\xi_N]/\mathfrak{p}$ is a field. In this case, $\dim(X) \in \mathbb{Z}[\xi_N]$ (see [61]) can be considered as an element in $\mathbb{Z}[\xi_N]/\mathfrak{p}$ in a natural way.

Theorem 5.4.2 *Let $\mathcal{C}$ be a spherical fusion category over $\mathbb{C}$ with the Frobenius-Schur exponent N. Let ξ_N be a primitive N-th root of unity with $\dim(X) \in \mathbb{Z}[\xi_N]$ for any object X of $\mathcal{C}$. For any prime ideal $\mathfrak{p}$ of $\mathbb{Z}[\xi_N]$, the Casimir number $m_{\mathcal{C}} \neq 0$ in $\mathbb{Z}[\xi_N]/\mathfrak{p}$ if and only if the global dimension $\dim(\mathcal{C}) \neq 0$ in $\mathbb{Z}[\xi_N]/\mathfrak{p}$.*

Proof Note that the Casimir number $m_{\mathcal{C}} = \sum_{i\in I} X_i a X_{i^*}$ for some $a \in \mathrm{Gr}(\mathcal{C})$. Applying dim to this equality, we have $m_{\mathcal{C}} = \dim(\mathcal{C})\dim(a)$. Thus, if $m_{\mathcal{C}} \neq 0$ in $\mathbb{Z}[\xi_N]/\mathfrak{p}$, then $\dim(\mathcal{C}) \neq 0$ in $\mathbb{Z}[\xi_N]/\mathfrak{p}$. Conversely, if $\dim(\mathcal{C}) \neq 0$ in $\mathbb{Z}[\xi_N]/\mathfrak{p}$, so is $\dim(\mathcal{C}) \neq 0$ in K, where K is an algebraic closure of the field $\mathbb{Z}[\xi_N]/\mathfrak{p}$. Let $Z(\mathcal{C})$ be the Drinfeld center of $\mathcal{C}$. Since $\dim(\mathcal{C}) \neq 0$ in K, it follows from [58, Section 5] that $Z(\mathcal{C})$ is a modular category and $\dim(Z(\mathcal{C})) = \dim(\mathcal{C})^2 \neq 0$ in K. We denote by $\mathrm{Irr}(Z(\mathcal{C}))$ the set of isomorphism classes of simple objects of $Z(\mathcal{C})$ and n the cardinality of $\mathrm{Irr}(Z(\mathcal{C}))$. By Example 5.1.8, the determinant of $Z(\mathcal{C})$ is

$$d_{Z(\mathcal{C})} = \frac{\dim(Z(\mathcal{C}))^n}{\prod\limits_{Y\in \mathrm{Irr}(Z(\mathcal{C}))} \dim(Y)^2} \neq 0.$$

Note that $d_{Z(\mathcal{C})}$ is the determinant of the matrix of left multiplication by the Casimir element $\sum_{Y\in\mathrm{Irr}(Z(\mathcal{C}))} YY^*$. It follows that $\sum_{Y\in\mathrm{Irr}(Z(\mathcal{C}))} YY^*$ is an invertible element in $\mathrm{Gr}(Z(\mathcal{C})) \otimes_{\mathbb{Z}} K$. Note that the forgetful tensor functor $F : Z(\mathcal{C}) \to \mathcal{C}$ induces an algebra homomorphism $f : \mathrm{Gr}(Z(\mathcal{C})) \otimes_{\mathbb{Z}} K \to \mathrm{Gr}(\mathcal{C}) \otimes_{\mathbb{Z}} K$ whose image is contained in the center of $\mathrm{Gr}(\mathcal{C}) \otimes_{\mathbb{Z}} K$. In particular, from the proof of [66, Lemma 3.1] we may see that

$$f(\sum_{Y\in\mathrm{Irr}(Z(\mathcal{C}))} YY^*) = \sum_{i,j\in I} X_i X_j X_{i^*} X_{j^*}.$$

Thus, $\sum_{i,j\in I} X_i X_j X_{i^*} X_{j^*}$ is a central invertible element in $\mathrm{Gr}(\mathcal{C}) \otimes_{\mathbb{Z}} K$. This together with Lemma 5.4.1 shows that the Schur element $c_V \neq 0$ for any simple $\mathrm{Gr}(\mathcal{C}) \otimes_{\mathbb{Z}} K$-module V. We conclude that $\mathrm{Gr}(\mathcal{C}) \otimes_{\mathbb{Z}} K$ is semisimple by [32, Theorem 7.2.6], and hence $m_{\mathcal{C}} \neq 0$ in K by Proposition 5.1.2. This completes the proof. □

We are now ready to state the relationship between the Casimir number $m_{\mathcal{C}}$ and the Frobenius-Schur exponent N of $\mathcal{C}$.

Theorem 5.4.3 *Let $\mathcal{C}$ be a spherical fusion category over $\mathbb{C}$. The Casimir number $m_{\mathcal{C}}$ and the Frobenius-Schur exponent N of $\mathcal{C}$ have the same prime factors.*

Proof For the Casimir number $m_{\mathcal{C}}$, there are some $a \in \mathrm{Gr}(\mathcal{C})$ such that

$$\sum_{i \in I} X_i a X_{i^*} = m_{\mathcal{C}}.$$

Applying dim to this equality, we have $\dim(\mathcal{C}) \dim(a) = m_{\mathcal{C}}$ in $\mathbb{Z}[\xi_N]$. Note that (N) and $(\dim(\mathcal{C}))$ are two principal ideals of $\mathbb{Z}[\xi_N]$ having the same prime ideal factors (see [10, Theorem 3.9]). If $p \mid N$ for a prime number p, then there exists a prime ideal factor $\mathfrak{p}$ of (N) such that $\mathfrak{p} \cap \mathbb{Z} = (p)$. In this case, $\mathfrak{p}$ is also a prime ideal factor of $(\dim(\mathcal{C}))$. Moreover, $(\dim(\mathcal{C})) \cap \mathbb{Z} \subseteq \mathfrak{p} \cap \mathbb{Z} = (p)$. It follows from $m_{\mathcal{C}} = \dim(\mathcal{C}) \dim(a)$ that $m_{\mathcal{C}} \in (\dim(\mathcal{C})) \cap \mathbb{Z} \subseteq (p)$, and hence $p \mid m_{\mathcal{C}}$. Conversely, if $p \nmid N$ for a prime p, we need to show that $p \nmid m_{\mathcal{C}}$. Let $\mathfrak{p}$ be a prime ideal of $\mathbb{Z}[\xi_N]$ such that $\mathfrak{p} \cap \mathbb{Z} = (p)$. Then $(N) \nsubseteq \mathfrak{p}$ since $p \nmid N$. This implies that $(\dim(\mathcal{C})) \nsubseteq \mathfrak{p}$, or $\dim(\mathcal{C}) \neq 0$ in $\mathbb{Z}[\xi_N]/\mathfrak{p}$. It follows from Theorem 5.4.2 that $m_{\mathcal{C}} \neq 0$ in $\mathbb{Z}[\xi_N]/\mathfrak{p}$. In other words, $m_{\mathcal{C}} \notin \mathfrak{p} \cap \mathbb{Z} = (p)$ and hence $p \nmid m_{\mathcal{C}}$. □

Remark 5.4.4 *Let (N) and $(\dim(\mathcal{C}))$ be principal ideals of $\mathbb{Z}[\xi_N]$ generated by N and $\dim(\mathcal{C})$ respectively. The statement of [10, Theorem 3.9] says that (N) and $(\dim(\mathcal{C}))$ have the same prime ideal factors. This is called the Cauchy theorem for a spherical fusion category. Indeed, applying this to the case $\mathcal{C}$ being $\mathbb{C}G$-mod for a finite group G, we obtain the classical Cauchy theorem for a finite group G, namely, $\dim(\mathcal{C}) = |G|$ and $N = \exp(G)$ have the same prime factors. Now Theorem 5.4.3 shows that the Casimir number $m_{\mathcal{C}}$ and the Frobenius-Schur exponent N of $\mathcal{C}$ have the same prime factors. This may be thought of as an integer version of the Cauchy theorem for a spherical fusion category.*

Chapter 6 Higher Frobenius-Schur Indicators in Positive Characteristic

Throughout this chapter, H is a finite dimensional semisimple Hopf algebra over an algebraically closed field $\Bbbk$ of characteristic p. We need to add the special condition that $S(\Lambda_{(2)})\Lambda_{(1)}$ is invertible in some sections so as to make sure that $\dim_{\Bbbk}(V) \neq 0$ for any simple H-module V. In this chapter, we generalize the notations of higher Frobenius-Schur indicators from characteristic 0 case to characteristic p case and find that the indicators defined here share some common properties with the ones defined over a field of characteristic 0. We show that the higher Frobenius-Schur indicators considered here are monoidal invariants of the representation category H-mod.

6.1 Characterizations of $S^2 = id$

In this section, we will give some sufficient and necessary conditions for $S^2 = id$. We need the following preparations.

Let H be a finite dimensional semisimple Hopf algebra over an algebraically closed field $\Bbbk$ of characteristic p. We fix a left integral Λ of H and a right integral λ of H^* such that $\lambda(\Lambda) = 1$. We denote $\{V_i \mid 0 \leqslant i \leqslant m-1\}$ the set of all simple left H-modules up to isomorphism and $\{e_i \mid 0 \leqslant i \leqslant m-1\}$ the set of all central primitive idempotents of H. The character of V_i is denoted by χ_i for $0 \leqslant i \leqslant m-1$ and the character of the left regular module H is denoted by χ_H. Obviously, $\chi_H = \sum_{i=0}^{m-1} \dim_{\Bbbk}(V_i)\chi_i$.

Recall that the dual Hopf algebra H^* has an H-bimodule structure given by

$$(a \rightharpoonup f)(b) = f(ba), \quad (f \leftharpoonup a)(b) = f(ab), \quad \text{for } a, b \in H,\ f \in H^*.$$

Moreover, $(H^*, \leftharpoonup)$ and $(\rightharpoonup, H^*)$ are both free H-modules generated by λ, i.e., $H^* = \lambda \leftharpoonup H$ and $H^* = H \rightharpoonup \lambda$ (see [70, Corollary 2(b)]). This provides an associative and non-degenerate bilinear form:

$$H \times H \to \Bbbk, \quad a \times b \mapsto \lambda(ab), \quad \text{for } a, b \in H.$$

Moreover, the pair (H, λ) is a Frobenius algebra with the Frobenius homomorphism

λ satisfying the equality (see [70, Eq.(1)]):

$$a = \lambda(a\Lambda_{(1)})S(\Lambda_{(2)}) = \lambda(S(\Lambda_{(2)})a)\Lambda_{(1)}, \quad \text{for } a \in H. \tag{6.1}$$

Since H is a semisimple Hopf algebra, the right integral $\lambda \in H^*$ satisfies $\lambda(ab) = \lambda(S^2(b)a)$ for $a, b \in H$ (see [70, Theorem 3(a)]). Thus, the Hopf algebra H is a symmetric Frobenius algebra whose symmetric bilinear form can be taken to be

$$H \times H \to \Bbbk, \ a \times b \mapsto \lambda(uab) = (\lambda \leftharpoonup u)(ab) = (u \rightharpoonup \lambda)(ab),$$

where u is a unit of H satisfying

$$S^2(h) = uhu^{-1}, \quad \text{for all } h \in H,$$

and the Frobenius homomorphism $\lambda \leftharpoonup u = u \rightharpoonup \lambda$ holds because

$$\lambda(au) = \lambda(S^2(u)a) = \lambda(ua), \quad \text{for all } a \in H.$$

We may see from [22, Lemma 1.4(2)] that the set $\{\Lambda_{(1)}, u^{-1}S(\Lambda_{(2)})\}$ is a pair of dual bases of H with respect to the Frobenius homomorphism $\lambda \leftharpoonup u$ $(= u \rightharpoonup \lambda)$. The symmetry of the Frobenius homomorphism $\lambda \leftharpoonup u$ $(= u \rightharpoonup \lambda)$ means that

$$\Lambda_{(1)} \otimes u^{-1}S(\Lambda_{(2)}) = u^{-1}S(\Lambda_{(2)}) \otimes \Lambda_{(1)}. \tag{6.2}$$

For any simple H-module V_i and any $\varphi \in \mathrm{End}_{\Bbbk}(V_i)$, we use the pair of dual bases $\{\Lambda_{(1)}, u^{-1}S(\Lambda_{(2)})\}$ with respect to the Frobenius homomorphism $\lambda \leftharpoonup u$ to define the map $\mathcal{I}(\varphi) \in \mathrm{End}_{\Bbbk}(V_i)$ by

$$\mathcal{I}(\varphi)(v) = \Lambda_{(1)}\varphi(u^{-1}S(\Lambda_{(2)})v), \quad \text{for } v \in V_i.$$

Note that $\mathcal{I}(\varphi)$ lies in $\mathrm{End}_H(V_i) \cong \Bbbk$. There exists a unique element $c_i \in \Bbbk$ such that

$$\mathcal{I}(\varphi) = c_i \mathrm{tr}(\varphi) id_{V_i}, \quad \text{for all } \varphi \in \mathrm{End}_k(V_i).$$

Such an element c_i, depending only on the isomorphism class of V_i, is called the *Schur element* associated to V_i (see [32, Theorem 7.2.1]). In particular, it follows from [32, Theorem 7.2.6] that the Schur element $c_i \neq 0$ in $\Bbbk$ for any simple H-module V_i and the Frobenius homomorphism $u \rightharpoonup \lambda$ of H can be written as follows:

$$\lambda \leftharpoonup u = u \rightharpoonup \lambda = \sum_{i=0}^{m-1} \frac{1}{c_i}\chi_i. \tag{6.3}$$

A relationship between the elements u and $\mathbf{u} := S(\Lambda_{(2)})\Lambda_{(1)}$ is given as follows:

Proposition 6.1.1 *With the notations above, we have*

$$\mathbf{u} = u \sum_{i=0}^{m-1} \dim_{\Bbbk}(V_i) c_i e_i.$$

Proof Note that each central primitive idempotent e_i acts as the identity on V_i and annihilates V_j for $j \neq i$. It follows that $\chi_j(e_i) = \dim_{\Bbbk}(V_i)\delta_{i,j}$. By (6.3) we have

$$\chi_i(a) = \chi_i(ae_i) = \sum_{j=0}^{m-1} \frac{1}{c_j} \chi_j(c_i a e_i) = (u \rightharpoonup \lambda)(c_i a e_i) = (u c_i e_i \rightharpoonup \lambda)(a).$$

Thus, $\chi_i = u c_i e_i \rightharpoonup \lambda$ and hence

$$\chi_H = \sum_{i=0}^{m-1} \dim_{\Bbbk}(V_i)\chi_i = u \sum_{i=0}^{m-1} \dim_{\Bbbk}(V_i) c_i e_i \rightharpoonup \lambda. \tag{6.4}$$

For any map $\varphi \in \mathrm{End}_{\Bbbk}(H)$, the trace of φ is $\mathrm{tr}(\varphi) = \lambda(\varphi(S(\Lambda_{(2)}))\Lambda_{(1)})$ (see [70, Theorem 2]). Taking into account that $\varphi = L_a$, where L_a is the left multiplication operator of H by a, we have

$$\chi_H(a) = \mathrm{tr}(L_a) = \lambda(aS(\Lambda_{(2)})\Lambda_{(1)}) = \lambda(a\mathbf{u}) = (\mathbf{u} \rightharpoonup \lambda)(a).$$

This implies that

$$\chi_H = \mathbf{u} \rightharpoonup \lambda. \tag{6.5}$$

Comparing (6.5) with (6.4) and using the non-degeneracy of the Frobenius homomorphism λ, we have

$$\mathbf{u} = u \sum_{i=0}^{m-1} \dim_{\Bbbk}(V_i) c_i e_i.$$

The proof is completed. □

Proposition 6.1.2 *Let H be a finite dimensional semisimple Hopf algebra over a field $\Bbbk$ of positive characteristic p. The following statements are equivalent:*

(1) *The element $\mathbf{u}$ is invertible.*

(2) *For any simple H-module V_i, $p \nmid \dim_{\Bbbk}(V_i)$.*

(3) *The regular character χ_H of H is non-degenerate, namely, if $\chi_H(ab) = 0$ for all $a \in H$, then $b = 0$.*

Proof (1) ⇔ (2). It follows from Proposition 6.1.1 that Part (1) and Part (2) are equivalent.

(1) ⇒ (3). Since λ is non-degenerate and $\mathbf{u}$ is invertible, it follows from (6.5) that χ_H is non-degenerate.

(3) ⇒ (2). It follows from $\chi_H = \sum_{i=0}^{m-1} \dim_{\Bbbk}(V_i)\chi_i$ that the non-degeneracy of χ_H implies that $p \nmid \dim_{\Bbbk}(V_i)$ for any simple H-module V_i. □

Remark 6.1.3 (1) *If $S^2 = id$, then*

$$\mathbf{u} = S(\Lambda_{(2)})\Lambda_{(1)} = S(\Lambda_{(2)})S^2(\Lambda_{(1)}) = S(S(\Lambda_{(1)})\Lambda_{(2)}) = \varepsilon(\Lambda) \neq 0.$$

Namely, $\mathbf{u}$ is a nonzero scalar in $\Bbbk$. Conversely, if $\mathbf{u}$ is a nonzero scalar, then u is central by Proposition 6.1.1. *It follows from $S^2(a) = uau^{-1}$ that $S^2 = id$.*

(2) *If $\mathbf{u}$ is invertible, then $\mathbf{u}$ and u are the same up to the central invertible element $\sum_{i=0}^{m-1} \dim_{\Bbbk}(V_i)c_ie_i$ by Proposition* 6.1.1. *Hence, $S^2(a) = uau^{-1}$ implies that $S^2(a) = \mathbf{u}a\mathbf{u}^{-1}$ for all $a \in H$.*

(3) *If the characteristic $p > \dim_{\Bbbk}(H)^{1/2}$, it follows that*

$$p^2 > \dim_{\Bbbk}(H) = \sum_{i=0}^{m-1} \dim_{\Bbbk}(V_i)^2 \geqslant \dim_{\Bbbk}(V_i)^2.$$

Hence $p \nmid \dim_{\Bbbk}(V_i)$ for $0 \leqslant i \leqslant m-1$. In this case, $\mathbf{u}$ is invertible by Proposition 6.1.2.

Let $g_0 \in G(H)$ and $\alpha \in \mathrm{Alg}(H, k)$ be the modular elements of H and H^* respectively. Recall that the Radford's formula of S^4 has the form (see [69, Proposition 6]):

$$S^4(a) = \alpha^{-1} \rightharpoonup (g_0 a g_0^{-1}) \leftharpoonup \alpha.$$

Since H is unimodular, i.e., $\alpha = \varepsilon$, the Radford's formula of S^4 now becomes

$$S^4(a) = g_0 a g_0^{-1}.$$

The distinguished group-like element g_0 and the integral Λ of H satisfy the following useful equality (see [70, Theorem 3(d)]):

$$\Lambda_{(2)} \otimes \Lambda_{(1)} = \Lambda_{(1)} \otimes S^2(\Lambda_{(2)})g_0. \tag{6.6}$$

We now give an equivalent condition for the equality $S^2 = id$ in terms of the integral Λ of H.

Theorem 6.1.4 *Let H be a finite dimensional semisimple Hopf algebra over a field $\Bbbk$ of positive characteristic p. The following statements are equivalent:*

(1) *The nonzero integral Λ of H is cocommutative.*

(2) *The nonzero integral λ of H^* is cocommutative.*

(3) *$S^2 = id$.*

Proof It can be seen from [70, Corollary 5] that Part (2) and Part (3) are equivalent. We only need to show that Part (1) and Part (3) are equivalent.

(1) $\Rightarrow$ (3). If Λ is cocommutative, then

$$\mathbf{u} = S(\Lambda_{(2)})\Lambda_{(1)} = S(\Lambda_{(1)})\Lambda_{(2)} = \varepsilon(\Lambda) \neq 0.$$

It follows from Remark 6.1.3 (1) that $S^2 = id$.

(3) $\Rightarrow$ (1). If $S^2 = id$, then $\mathbf{u} = \varepsilon(\Lambda) \neq 0$. Applying $S \otimes id$ to both sides of the equality (6.6) and multiplying the tensor factors together, we have $\mathbf{u} = S(\mathbf{u})g_0$. Using $\mathbf{u} = \varepsilon(\Lambda)$ we have $g_0 = 1$. Now $g_0 = 1$ and $S^2 = id$, the equality (6.6) becomes $\Lambda_{(2)} \otimes \Lambda_{(1)} = \Lambda_{(1)} \otimes \Lambda_{(2)}$. We complete the proof. □

6.2 Some properties of the element u

We assume in this section that $\mathbf{u} = S(\Lambda_{(2)})\Lambda_{(1)}$ is an invertible element of H. We then describe some properties of the element $\mathbf{u}$. By Proposition 6.1.2 and Remark 6.1.3 (2), we have

- $S^2(a) = \mathbf{u}a\mathbf{u}^{-1}$ for all $a \in H$;
- $\dim_{\Bbbk}(V_i) \neq 0$ in $\Bbbk$ for $0 \leqslant i \leqslant m-1$.

Now we replace u with $\mathbf{u}$ in the equality (6.2). It turns out that

$$\Lambda_{(1)} \otimes \mathbf{u}^{-1}S(\Lambda_{(2)}) = \mathbf{u}^{-1}S(\Lambda_{(2)}) \otimes \Lambda_{(1)}, \tag{6.7}$$

which forms a pair of dual bases of H with respect to the Frobenius homomorphism $\lambda \leftharpoonup \mathbf{u}$. The Schur element associated to the simple H-module V_i under the pair of dual bases $\{\Lambda_{(1)}, \mathbf{u}^{-1}S(\Lambda_{(2)})\}$ with respect to the Frobenius homomorphism $\lambda \leftharpoonup \mathbf{u}$ is $1/\dim_{\Bbbk}(V_i)$. The equality (6.3) turns out to be

$$\lambda \leftharpoonup \mathbf{u} = \mathbf{u} \rightharpoonup \lambda = \sum_{i=0}^{m-1} \dim_{\Bbbk}(V_i)\chi_i = \chi_H. \tag{6.8}$$

By applying [22, Theorem 1.5] and (6.7), we obtain the expression of each central primitive idempotent e_i of H as follows:

$$e_i = \dim_{\Bbbk}(V_i)\chi_i(\Lambda_{(1)})\mathbf{u}^{-1}S(\Lambda_{(2)}) = \dim_{\Bbbk}(V_i)\chi_i(\mathbf{u}^{-1}S(\Lambda_{(2)}))\Lambda_{(1)}. \tag{6.9}$$

With these preparations, we can give some properties of the element $\mathbf{u}$ as follows:

Proposition 6.2.1 *The element* $\mathbf{u} = S(\Lambda_{(2)})\Lambda_{(1)}$ *satisfies the following properties*:

(1) $\mathbf{u} = \chi_H(\Lambda_{(1)})S(\Lambda_{(2)})$.

(2) $\Lambda_{(1)}\mathbf{u}^{-1}S(\Lambda_{(2)}) = 1$.

(3) $\lambda(e_i) = \dim_{\Bbbk}(V_i)\chi_i(\mathbf{u}^{-1})$.

(4) $\mathbf{u}S(\mathbf{u}) = S(\mathbf{u})\mathbf{u} = \varepsilon(\Lambda)\sum_{i=0}^{m-1} \dfrac{\dim_{\Bbbk}(V_i)^2}{\lambda(e_i)}e_i$.

(5) $S(\mathbf{u}^{-1})\mathbf{u} = \mathbf{u}S(\mathbf{u}^{-1}) = g_0$, *the distinguished group-like element of* H.

Proof (1) It follows from (6.9) that

$$e_i\mathbf{u} = \dim_{\Bbbk}(V_i)\chi_i(\Lambda_{(1)})S(\Lambda_{(2)}).$$

Thus,

$$\mathbf{u}=\sum_{i=0}^{m-1}e_i\mathbf{u}=\sum_{i=0}^{m-1}\dim_{\Bbbk}(V_i)\chi_i(\Lambda_{(1)})S(\Lambda_{(2)})=\chi_H(\Lambda_{(1)})S(\Lambda_{(2)}).$$

(2) Since

$$\Lambda_{(1)}\otimes\mathbf{u}^{-1}S(\Lambda_{(2)})=\mathbf{u}^{-1}S(\Lambda_{(2)})\otimes\Lambda_{(1)},$$

by (6.7), we obtain the desired result by multiplying the tensor factors together.

(3) Since

$$e_i=\dim_{\Bbbk}(V_i)\chi_i(\Lambda_{(1)})\mathbf{u}^{-1}S(\Lambda_{(2)}),$$

it follows that

$$e_i=\mathbf{u}e_i\mathbf{u}^{-1}=\dim_{\Bbbk}(V_i)\chi_i(\Lambda_{(1)})S(\Lambda_{(2)})\mathbf{u}^{-1}.$$

Hence

$$\lambda(e_i)=\dim_{\Bbbk}(V_i)\chi_i(\Lambda_{(1)})\lambda(S(\Lambda_{(2)})\mathbf{u}^{-1})=\dim_{\Bbbk}(V_i)\chi_i(\mathbf{u}^{-1}),$$

where the last equality follows from (6.1).

(4) For any $a\in H$, we have

$$S^3(a)=S(S^2(a))=S(\mathbf{u}a\mathbf{u}^{-1})=S(\mathbf{u}^{-1})S(a)S(\mathbf{u}),$$

we also have

$$S^3(a)=S^2(S(a))=\mathbf{u}S(a)\mathbf{u}^{-1}.$$

It follows that $S(\mathbf{u})\mathbf{u}$ is a central unit of H. The equality $\mathbf{u}S(\mathbf{u})=S(\mathbf{u})\mathbf{u}$ holds because

$$S(\mathbf{u})=S(S^2(\mathbf{u}))=S^2(S(\mathbf{u}))=\mathbf{u}S(\mathbf{u})\mathbf{u}^{-1}.$$

For the central unit $\mathbf{u}S(\mathbf{u})$, we suppose that

$$\mathbf{u}S(\mathbf{u})=\sum_{i=0}^{m-1}k_ie_i.$$

It is obvious that each scalar $k_i\neq 0$, then $e_i\mathbf{u}^{-1}=k_i^{-1}e_iS(\mathbf{u})$. We have

$$\begin{aligned}\lambda(e_i)&=(\mathbf{u}^{-1}\rightharpoonup\chi_H)(e_i)=\chi_H(e_i\mathbf{u}^{-1})=\frac{1}{k_i}\chi_H(e_iS(\mathbf{u}))\\&=\frac{\dim_{\Bbbk}(V_i)}{k_i}\chi_i(e_iS(\mathbf{u}))=\frac{\dim_{\Bbbk}(V_i)}{k_i}\chi_i(S(\mathbf{u}))\\&=\frac{\dim_{\Bbbk}(V_i)}{k_i}(\chi_i\circ S)(\mathbf{u})=\frac{\dim_{\Bbbk}(V_i)}{k_i}(\chi_i\circ S)(S(\Lambda_{(2)})\Lambda_{(1)})\end{aligned}$$

$$= \frac{\dim_{\Bbbk}(V_i)}{k_i}(\chi_i \circ S)(\Lambda_{(1)}S(\Lambda_{(2)})) = \frac{\dim_{\Bbbk}(V_i)^2\varepsilon(\Lambda)}{k_i} \neq 0.$$

It follows that $k_i = \dfrac{\dim_{\Bbbk}(V_i)^2\varepsilon(\Lambda)}{\lambda(e_i)}$ and hence

$$\mathbf{u}S(\mathbf{u}) = \sum_{i=0}^{m-1} k_i e_i = \varepsilon(\Lambda)\sum_{i=0}^{m-1}\frac{\dim_{\Bbbk}(V_i)^2}{\lambda(e_i)}e_i.$$

(5) Applying $S \otimes id$ to both sides of the equality (6.6) and multiplying the tensor factors together, we have $\mathbf{u} = S(\mathbf{u})g_0$ or $g_0 = S(\mathbf{u}^{-1})\mathbf{u}$. The equality $S(\mathbf{u}^{-1})\mathbf{u} = \mathbf{u}S(\mathbf{u}^{-1})$ follows from Part (4). □

As a consequence, we obtain the following result:

Corollary 6.2.2 *For any central primitive idempotent e_i of H, we have $\lambda(e_i) = \lambda(S(e_i))$.*

Proof There is a permutation $*$ on the index set $\{0, 1, \cdots, m-1\}$ determined by $i^* = j$ if $V_i^* \cong V_j$. Note that $S(e_i) = e_{i^*}$, $\dim_{\Bbbk}(V_i) = \dim_{\Bbbk}(V_{i^*})$ and $\chi_i \circ S = \chi_{i^*}$ (see [22, Lemma 1.8]). By Proposition 6.2.1 (3) we have

$$\lambda(S(e_i)) = \lambda(e_{i^*}) = \dim_{\Bbbk}(V_{i^*})\chi_{i^*}(\mathbf{u}^{-1}) = \dim_{\Bbbk}(V_i)\chi_i(S(\mathbf{u}^{-1})).$$

Since

$$S(\mathbf{u}^{-1}) = \mathbf{u}\frac{1}{\varepsilon(\Lambda)}\sum_{i=0}^{m-1}\frac{\lambda(e_i)}{\dim_{\Bbbk}(V_i)^2}e_i$$

by Proposition 6.2.1 (4), it follows that

$$\begin{aligned}\lambda(S(e_i)) &= \dim_{\Bbbk}(V_i)\chi_i(S(\mathbf{u}^{-1})) = \frac{\lambda(e_i)}{\varepsilon(\Lambda)\dim_{\Bbbk}(V_i)}\chi_i(\mathbf{u}) \\ &= \frac{\lambda(e_i)}{\varepsilon(\Lambda)\dim_{\Bbbk}(V_i)}\chi_i(S(\Lambda_{(2)})\Lambda_{(1)}) = \frac{\lambda(e_i)}{\varepsilon(\Lambda)\dim_{\Bbbk}(V_i)}\chi_i(\Lambda_{(1)}S(\Lambda_{(2)})) \\ &= \lambda(e_i).\end{aligned}$$

We complete the proof. □

Remark 6.2.3 *In summary, the permutation $*$ on the index set $\{0, 1, \cdots, m-1\}$ satisfies that*

$$i^{**} = i, \quad S(e_i) = e_{i^*}, \quad \dim_{\Bbbk}(V_{i^*}) = \dim_{\Bbbk}(V_i),$$

and by Corollary 6.2.2 that

$$\lambda(e_{i^*}) = \lambda(e_i), \quad \textit{for } 0 \leqslant i \leqslant m-1.$$

6.3 Higher Frobenius-Schur indicators

If $\mathrm{char}(\Bbbk) = 0$, the n-th Frobenius-Schur indicators of finite dimensional representations of semisimple Hopf algebras have been studied in [43]. In this section, we assume that $\mathbf{u} = S(\Lambda_{(2)})\Lambda_{(1)}$ is invertible, we will generalize these indicators from characteristic 0 to the case of characteristic p and describe them via a nonzero integral Λ of H. We begin with the following preparations.

Let H be a finite dimensional semisimple Hopf algebra over a field $\Bbbk$ of characteristic p with a nonzero integral Λ. Suppose $\mathbf{u} = S(\Lambda_{(2)})\Lambda_{(1)}$ is invertible. Applying $\Delta_{n-1} \otimes id$ to both sides of the equality:

$$\Lambda_{(2)} \otimes \Lambda_{(1)} = \Lambda_{(1)} \otimes S^2(\Lambda_{(2)})g_0 \quad \text{(see (6.6))},$$

we have

$$\Lambda_{(2)} \otimes \cdots \otimes \Lambda_{(n)} \otimes \Lambda_{(1)} = \Lambda_{(1)} \otimes \cdots \otimes \Lambda_{(n-1)} \otimes S^2(\Lambda_{(n)})g_0.$$

Since

$$g_0 = \mathbf{u}S(\mathbf{u}^{-1}) \quad \text{and} \quad S^2(\Lambda_{(n)}) = \mathbf{u}\Lambda_{(n)}\mathbf{u}^{-1},$$

the above equality induces the following equality:

$$\Lambda_{(2)} \otimes \cdots \otimes \Lambda_{(n)} \otimes \mathbf{u}^{-1}\Lambda_{(1)} = \Lambda_{(1)} \otimes \cdots \otimes \Lambda_{(n-1)} \otimes \Lambda_{(n)}S(\mathbf{u}^{-1}). \tag{6.10}$$

Note that the category H-mod of finite dimensional left H-modules is a semisimple monoidal category. Let $j_{\mathbf{u}} : id \to (-)^{**}$ be a natural isomorphism between the identity functor and the functor of taking the second dual. It is completely determined by a collection of H-module isomorphisms

$$j_{\mathbf{u},V} : V \to V^{**}, \quad j_{\mathbf{u},V}(v)(\vartheta) = \vartheta(\mathbf{u}v), \quad \text{for } v \in V, \vartheta \in V^*.$$

The inverse of $j_{\mathbf{u},V}$ is

$$j_{\mathbf{u},V}^{-1} : V^{**} \to V, \quad \alpha \mapsto j_{\mathbf{u},V}^{-1}(\alpha),$$

where $j_{\mathbf{u},V}^{-1}(\alpha) \in V$ satisfies the equality

$$\vartheta(j_{\mathbf{u},V}^{-1}(\alpha)) = \alpha(S^{-1}(\mathbf{u}^{-1})\vartheta), \quad \text{for } \vartheta \in V^*.$$

Since $S^2(a) = \mathbf{u}a\mathbf{u}^{-1}$ and $\mathbf{u}$ is not known to be a group-like element, the natural isomorphism $j_{\mathbf{u}}$ is not necessarily a tensor isomorphism. Although the representation category H-mod with respect to the structure $j_{\mathbf{u}}$ is not necessarily pivotal, we may still define higher Frobenius-Schur indicators for any finite dimensional H-module using the structure $j_{\mathbf{u}}$ of H-mod.

We denote $V^{\otimes n}$ the n-th tensor power of V where $V^{\otimes 0}$ is the trivial H-module $\Bbbk$. For any natural number $n \geqslant 1$, we define the following $\Bbbk$-linear map:

$$E_V^n : \mathrm{Hom}_H(\Bbbk, V^{\otimes n}) \to \mathrm{Hom}_H(\Bbbk, V^{\otimes n}), \quad f \mapsto E_V^n(f),$$

where $E_V^n(f)$ is an H-module morphism from $\Bbbk$ to $V^{\otimes n}$ given by

$$E_V^n(f) : \Bbbk \xrightarrow{\mathrm{coev}_{V^*}} V^* \otimes V^{**} = V^* \otimes \Bbbk \otimes V^{**} \xrightarrow{id \otimes f \otimes id} V^* \otimes V^{\otimes n} \otimes V^{**}$$
$$\xrightarrow{\mathrm{ev}_V \otimes id} V^{\otimes(n-1)} \otimes V^{**} \xrightarrow{id \otimes j_{\mathbf{u},V}^{-1}} V^{\otimes n}.$$

Here the maps coev_{V^*} and ev_V are the usual coevaluation morphism of V^* and evaluation morphism of V respectively. If we set $f(1) = \sum v_1 \otimes \cdots \otimes v_n \in V^{\otimes n}$, the above definition of $E_V^n(f)$ shows that

$$E_V^n(f)(1) = \sum v_2 \otimes \cdots \otimes v_n \otimes \mathbf{u}^{-1} v_1. \tag{6.11}$$

Similarly to [62], we give the definition of the n-th Frobenius-Schur indicator of V to be the trace of the linear operator E_V^n as follows:

Definition 6.3.1 *Let H be a finite dimensional semisimple Hopf algebra over a field $\Bbbk$ of characteristic p with $\mathbf{u} = S(\Lambda_{(2)})\Lambda_{(1)}$ being invertible. For any finite dimensional H-module V, the n-th Frobenius-Schur indicator of V is defined by*

$$\nu_n(V) = \mathrm{tr}(E_V^n), \quad \text{for } n \geqslant 1.$$

Similarly to the characteristic 0 case, the n-th Frobenius-Schur indicator of V defined above can also be described by a nonzero integral Λ of H:

Theorem 6.3.2 *Let Λ be a nonzero integral of H with $\mathbf{u} = S(\Lambda_{(2)})\Lambda_{(1)}$ being invertible. Suppose χ_V is the character of a finite dimensional H-module V. We have*

$$\nu_n(V) = \chi_V(\mathbf{u}^{-1}\Lambda_{(1)} \cdots \Lambda_{(n)}), \quad \text{for } n \geqslant 1.$$

Proof We first show that the equality $\nu_n(V) = \chi_V(\mathbf{u}^{-1}\Lambda_{(1)} \cdots \Lambda_{(n)})$ holds for an idempotent integral Λ. Suppose that α is the following $\Bbbk$-linear map:

$$\alpha : V^{\otimes n} \to V^{\otimes n}, \quad v_1 \otimes v_2 \otimes \cdots \otimes v_n \mapsto v_2 \otimes \cdots \otimes v_n \otimes v_1$$

and

$$\delta = \alpha \circ (\mathbf{u}^{-1}\Lambda_{(1)} \otimes \Lambda_{(2)} \otimes \cdots \otimes \Lambda_{(n)}).$$

We have

$$\delta(v_1 \otimes v_2 \otimes \cdots \otimes v_n) = \alpha(\mathbf{u}^{-1}\Lambda_{(1)} v_1 \otimes \Lambda_{(2)} v_2 \otimes \cdots \otimes \Lambda_{(n)} v_n)$$

$$\begin{aligned}
&= \Lambda_{(2)}v_2 \otimes \cdots \otimes \Lambda_{(n)}v_n \otimes \mathbf{u}^{-1}\Lambda_{(1)}v_1 \\
&= \Lambda_{(1)}v_2 \otimes \cdots \otimes \Lambda_{(n-1)}v_n \otimes \Lambda_{(n)}S(\mathbf{u}^{-1})v_1 \quad \text{(by (6.10))} \\
&= \Lambda \cdot (v_2 \otimes \cdots \otimes v_n \otimes S(\mathbf{u}^{-1})v_1). \qquad (6.12)
\end{aligned}$$

This shows that

$$\delta(V^{\otimes n}) \subseteq \Lambda \cdot V^{\otimes n} = (V^{\otimes n})^H.$$

Note that the map

$$\Phi : \mathrm{Hom}_H(\Bbbk, V^{\otimes n}) \to (V^{\otimes n})^H, \quad f \mapsto f(1)$$

is an H-module isomorphism. We claim that the following diagram is commutative:

$$\begin{array}{ccc}
\mathrm{Hom}_H(\Bbbk, V^{\otimes n}) & \xrightarrow{E_V^n} & \mathrm{Hom}_H(\Bbbk, V^{\otimes n}) \\
\Phi \downarrow & & \downarrow \Phi \\
(V^{\otimes n})^H & \xrightarrow{\quad \delta \quad} & (V^{\otimes n})^H
\end{array}$$

Indeed, for any $f \in \mathrm{Hom}_H(\Bbbk, V^{\otimes n})$, we suppose that

$$f(1) = \sum v_1 \otimes \cdots \otimes v_n \in V^{\otimes n}.$$

It follows from $f(1) = f(\Lambda \cdot 1) = \Lambda \cdot f(1)$ that

$$\sum v_1 \otimes \cdots \otimes v_n = \sum \Lambda_{(1)}v_1 \otimes \cdots \otimes \Lambda_{(n)}v_n. \qquad (6.13)$$

On the one hand, we have

$$\begin{aligned}
(\delta \circ \Phi)(f) &= \delta(f(1)) = \delta(\sum v_1 \otimes \cdots \otimes v_n) \\
&= \Lambda \cdot (\sum v_2 \otimes \cdots \otimes v_n \otimes S(\mathbf{u}^{-1})v_1) \quad \text{(by (6.12))},
\end{aligned}$$

On the other hand, we have

$$\begin{aligned}
(\Phi \circ E_V^n)(f) &= E_V^n(f)(1) = \sum v_2 \otimes \cdots \otimes v_n \otimes \mathbf{u}^{-1}v_1 \quad \text{(by (6.11))} \\
&= \sum \Lambda_{(2)}v_2 \otimes \cdots \otimes \Lambda_{(n)}v_n \otimes \mathbf{u}^{-1}\Lambda_{(1)}v_1 \quad \text{(by (6.13))} \\
&= \sum \Lambda_{(1)}v_2 \otimes \cdots \otimes \Lambda_{(n-1)}v_n \otimes \Lambda_{(n)}S(\mathbf{u}^{-1})v_1 \quad \text{(by (6.10))} \\
&= \Lambda \cdot (\sum v_2 \otimes \cdots \otimes v_n \otimes S(\mathbf{u}^{-1})v_1).
\end{aligned}$$

We obtain that $\delta \circ \Phi = \Phi \circ E_V^n$, or equivalently, $E_V^n = \Phi^{-1} \circ \delta \circ \Phi$. It follows that

$$\begin{aligned}\nu_n(V) &= \mathrm{tr}(E_V^n) = \mathrm{tr}_{V^{\otimes n}}(\delta)\\ &= \mathrm{tr}_{V^{\otimes n}}(\alpha \circ (\mathbf{u}^{-1}\Lambda_{(1)} \otimes \Lambda_{(2)} \otimes \cdots \otimes \Lambda_{(n)}))\\ &= \mathrm{tr}_V(\mathbf{u}^{-1}\Lambda_{(1)} \cdots \Lambda_{(n)})\\ &= \chi_V(\mathbf{u}^{-1}\Lambda_{(1)} \cdots \Lambda_{(n)}),\end{aligned}$$

where the equality

$$\mathrm{tr}_{V^{\otimes n}}(\alpha \circ (\mathbf{u}^{-1}\Lambda_{(1)} \otimes \Lambda_{(2)} \otimes \cdots \otimes \Lambda_{(n)})) = \mathrm{tr}_V(\mathbf{u}^{-1}\Lambda_{(1)} \cdots \Lambda_{(n)})$$

follows from [43, Lemma 2.3]. We have shown that

$$\nu_n(V) = \chi_V(\mathbf{u}^{-1}\Lambda_{(1)} \cdots \Lambda_{(n)}),$$

where Λ is idempotent. Since $\mathbf{u}^{-1}\Lambda_{(1)} \cdots \Lambda_{(n)}$ does not depend on the choice of a nonzero integral Λ, the equality $\nu_n(V) = \chi_V(\mathbf{u}^{-1}\Lambda_{(1)} \cdots \Lambda_{(n)})$ holds for any nonzero integral Λ of H. □

Remark 6.3.3 *If* $\mathrm{char}(\Bbbk) = 0$ *and Λ is idempotent, then* $\mathbf{u} = \varepsilon(\Lambda) = 1$. *In this case, the n-th Frobenius-Schur indicator of V is* $\chi_V(\Lambda_{(1)} \cdots \Lambda_{(n)})$, *which is the one defined in* [43, *Definition* 2.3].

In the rest of this section, we will extend the n-th Frobenius-Schur indicator $\nu_n(V)$ of V from $n \geqslant 1$ to the case $n \in \mathbb{Z}$. Recall that the n-th Sweedler power map $P_n : H \to H$ is defined by

$$P_n(a) = \begin{cases} a_{(1)} \cdots a_{(n)}, & n \geqslant 1,\\ \varepsilon(a), & n = 0,\\ S(a_{(1)}) \cdots S(a_{(-n)}), & n \leqslant -1.\end{cases}$$

From the n-th Sweedler power map P_n of H, we may see that

$$\nu_n(V) = \chi_V(\mathbf{u}^{-1}P_n(\Lambda)), \quad \text{for } n \geqslant 1.$$

However, this expression is well-defined for any integer n. Thus, we may extend this formula from $n \geqslant 1$ to any integer n.

Definition 6.3.4 *Let H be a finite dimensional semisimple Hopf algebra over a field $\Bbbk$ of characteristic p with* $\mathbf{u} = S(\Lambda_{(2)})\Lambda_{(1)}$ *being invertible. For any finite dimensional H-module V and any $n \in \mathbb{Z}$, the n-th Frobenius-Schur indicator of V is defined by*

$$\nu_n(V) = \chi_V(\mathbf{u}^{-1}P_n(\Lambda)).$$

Remark 6.3.5 (1) *Note that* $S(\Lambda) = \Lambda$. *The n-th Frobenius-Schur indicator of V can be written as*

$$\nu_n(V) = \begin{cases} \chi_V(\mathbf{u}^{-1}\Lambda_{(1)} \cdots \Lambda_{(n)}), & n \geqslant 1, \\ \chi_V(\mathbf{u}^{-1}\varepsilon(\Lambda)), & n = 0, \\ \chi_V(\mathbf{u}^{-1}\Lambda_{(-n)} \cdots \Lambda_{(1)}), & n \leqslant -1. \end{cases}$$

(2) *By Proposition* 6.2.1 (4), *we have*

$$\mathbf{u}^{-1}S(\mathbf{u}^{-1}) = \sum_{i=0}^{m-1} \frac{\lambda(e_i)}{\varepsilon(\Lambda)\dim_{\Bbbk}(V_i)^2} e_i \in Z(H).$$

It follows that

$$\begin{aligned} \nu_0(V) &= \varepsilon(\Lambda)\chi_V(\mathbf{u}^{-1}) = \varepsilon(\Lambda)\chi_V(\mathbf{u}^{-1}S(\mathbf{u}^{-1})S(\mathbf{u})) \\ &= \sum_{i=0}^{m-1} \frac{\lambda(e_i)}{\dim_{\Bbbk}(V_i)^2}\chi_V(e_iS(\mathbf{u})) = \sum_{i=0}^{m-1} \frac{\lambda(e_i)}{\dim_{\Bbbk}(V_i)^2}\chi_V(e_iS(\Lambda_{(1)})S^2(\Lambda_{(2)})) \\ &= \sum_{i=0}^{m-1} \frac{\lambda(e_i)}{\dim_{\Bbbk}(V_i)^2}\chi_V(e_iS^2(\Lambda_{(2)})S(\Lambda_{(1)})) = \varepsilon(\Lambda)\sum_{i=0}^{m-1} \frac{\lambda(e_i)}{\dim_{\Bbbk}(V_i)^2}\chi_V(e_i). \end{aligned}$$

(3) $\nu_{-1}(V) = \nu_1(V) = \chi_V(\mathbf{u}^{-1}\Lambda) = \chi_V\left(\dfrac{\Lambda}{\varepsilon(\Lambda)}\right)$.

(4) *By* [91, *Proposition* 3.1], $\Lambda_{(1)}\Lambda_{(2)}$ *and* $\Lambda_{(2)}\Lambda_{(1)}$ *are both central elements of H, they are determined by the values that the characters* χ_i *for all* $0 \leqslant i \leqslant m-1$ *take on them. It follows from* $\chi_i(\Lambda_{(1)}\Lambda_{(2)}) = \chi_i(\Lambda_{(2)}\Lambda_{(1)})$ *that* $\Lambda_{(1)}\Lambda_{(2)} = \Lambda_{(2)}\Lambda_{(1)}$. *Therefore,* $\nu_{-2}(V) = \nu_2(V)$.

The higher Frobenius-Schur indicators of any simple H-module V_i can be described as follows:

Proposition 6.3.6 *For any* $n \in \mathbb{Z}$ *and any simple* H*-module* V_i *with the character* χ_i, *we have*

$$\nu_n(V_i) = \frac{\chi_i(P_n(\Lambda))\lambda(e_i)}{\dim_{\Bbbk}(V_i)^2}.$$

Proof Since $P_n(\Lambda) \in Z(H)$ for any $n \in \mathbb{Z}$ (see [91, Proposition 3.1]), it follows that

$$P_n(\Lambda) = \sum_{i=0}^{m-1} \frac{\chi_i(P_n(\Lambda))}{\dim_{\Bbbk}(V_i)} e_i.$$

The n-th Frobenius-Schur indicator of V_i is

$$\nu_n(V_i) = \chi_i(\mathbf{u}^{-1}P_n(\Lambda)) = \frac{\chi_i(P_n(\Lambda))}{\dim_{\Bbbk}(V_i)}\chi_i(\mathbf{u}^{-1}) = \frac{\chi_i(P_n(\Lambda))\lambda(e_i)}{\dim_{\Bbbk}(V_i)^2},$$

where the last equality follows from Proposition 6.2.1 (3). □

For any semisimple Hopf algebra H over a field $\Bbbk$ of characteristic 0, the finite dimensional H-module V and its dual V^* have the same n-th Frobenius-Schur indicators for all $n \geqslant 1$ (see [43, Section 2.3]). The following result shows that this also holds for the n-th Frobenius-Schur indicators defined for the Hopf algebra over a field $\Bbbk$ of characteristic p.

Proposition 6.3.7 *Let H be a finite dimensional semisimple Hopf algebra over a field $\Bbbk$ of characteristic p with $\mathbf{u} = S(\Lambda_{(2)})\Lambda_{(1)}$ being invertible. Let V be a finite dimensional H-module and V^* the dual module of V. We have $\nu_n(V) = \nu_n(V^*)$ for all $n \in \mathbb{Z}$.*

Proof Since $S(\Lambda) = \Lambda$, we have

$$S(P_n(\Lambda)) = P_n(\Lambda), \quad \text{for } n \in \mathbb{Z}.$$

For the case $n \geqslant 1$, the n-th Frobenius-Schur indicator of V^* is

$$\begin{aligned}\nu_n(V^*) &= (\chi_{V^*})(\mathbf{u}^{-1}P_n(\Lambda)) = (\chi_V \circ S)(\mathbf{u}^{-1}P_n(\Lambda))\\ &= \chi_V(\Lambda_{(1)}\cdots\Lambda_{(n)}S(\mathbf{u}^{-1})) = \chi_V(\Lambda_{(2)}\cdots\Lambda_{(n)}\mathbf{u}^{-1}\Lambda_{(1)}) \quad \text{(by (6.10))}\\ &= \chi_V(\mathbf{u}^{-1}\Lambda_{(1)}\Lambda_{(2)}\cdots\Lambda_{(n)}) = \nu_n(V).\end{aligned}$$

For the case $n \leqslant -1$, the n-th Frobenius-Schur indicator of V^* is

$$\begin{aligned}\nu_n(V^*) &= (\chi_{V^*})(\mathbf{u}^{-1}P_n(\Lambda)) = (\chi_V \circ S)(\mathbf{u}^{-1}P_n(\Lambda))\\ &= \chi_V(\Lambda_{(-n)}\cdots\Lambda_{(1)}S(\mathbf{u}^{-1})) = \chi_V(S(\mathbf{u}^{-1})\Lambda_{(-n)}\cdots\Lambda_{(1)})\\ &= \chi_V(S(\mathbf{u}^{-1})\mathbf{u}^{-1}\Lambda_{(1)}S(\mathbf{u})\Lambda_{(-n)}\cdots\Lambda_{(2)}) \quad \text{(by (6.10))}\\ &= \chi_V(\Lambda_{(1)}\mathbf{u}^{-1}\Lambda_{(-n)}\cdots\Lambda_{(2)}) = \chi_V(\mathbf{u}^{-1}\Lambda_{(-n)}\cdots\Lambda_{(1)})\\ &= \nu_n(V).\end{aligned}$$

For the case $n = 0$, we denote $S(e_i) = e_{i^*}$ for $0 \leqslant i \leqslant m-1$, then $*$ is a permutation of $\{0, 1, \cdots, m-1\}$, $V_{i^*} \cong V_i^*$ and $\lambda(e_{i^*}) = \lambda(e_i)$ by Corollary 6.2.2. We have

$$\begin{aligned}\nu_0(V^*) &= \varepsilon(\Lambda)\sum_{i=0}^{m-1}\frac{\lambda(e_i)}{\dim_{\Bbbk}(V_i)^2}\chi_V(S(e_i)) \quad \text{(by Remark 6.3.5 (2))}\\ &= \varepsilon(\Lambda)\sum_{i=0}^{m-1}\frac{\lambda(e_{i^*})}{\dim_{\Bbbk}(V_{i^*})^2}\chi_V(e_{i^*})\\ &= \varepsilon(\Lambda)\sum_{i=0}^{m-1}\frac{\lambda(e_i)}{\dim_{\Bbbk}(V_i)^2}\chi_V(e_i)\end{aligned}$$

$$= \nu_0(V).$$

The proof is completed. □

Kashina, Sommerhäuser and Zhu have shown in [43, Proposition 2.5] that the n-th Frobenius-Schur indictor of the regular module of a semisimple Hopf algebra over a field of characteristic 0 can be described as $\mathrm{tr}(S \circ P_{n-1})$ for $n \geqslant 1$. The following result shows that this formula also holds for the n-th Frobenius-Schur indicator defined for the Hopf algebra H over a field $\Bbbk$ of characteristic p.

Proposition 6.3.8 *Let H be a finite dimensional semisimple Hopf algebra over a field $\Bbbk$ of characteristic p with $\mathbf{u} = S(\Lambda_{(2)})\Lambda_{(1)}$ being invertible. For any $n \in \mathbb{Z}$, the n-th Frobenius-Schur indictor of the regular module of H can be written as $\nu_n(H) = \mathrm{tr}(S \circ P_{n-1})$, where P_{n-1} is the $(n-1)$-th Sweedler power map of H.*

Proof Note that the left integral Λ of H and the right integral λ of H^* satisfy $\lambda(\Lambda) = 1$. For any $n \in \mathbb{Z}$, by Radford's trace formula [70, Theorem 2], we have

$$\begin{aligned}\mathrm{tr}(S \circ P_{n-1}) &= \mathrm{tr}(P_{n-1} \circ S) = \lambda(S(\Lambda_{(2)})(P_{n-1} \circ S)(\Lambda_{(1)}))\\ &= \lambda(S(\Lambda_{(2)})P_{n-1}(S(\Lambda_{(1)}))) = \lambda(\Lambda_{(1)}P_{n-1}(\Lambda_{(2)}))\\ &= \lambda(P_n(\Lambda)) = \chi_H(\mathbf{u}^{-1}P_n(\Lambda)) \ \text{ by (6.8)}\\ &= \nu_n(H).\end{aligned}$$

The proof is completed. □

6.4 Monoidal invariants

In this section, we will show that the n-th Frobenius-Schur indicator $\nu_n(V)$ defined in previous section is a monoidal invariant of the representation category H-mod for any $n \in \mathbb{Z}$.

Recall from [1] that a (normalized) twist for semisimple Hopf algebra H is an invertible element $J \in H \otimes H$ that satisfies

$$(\varepsilon \otimes id)(J) = (id \otimes \varepsilon)(J) = 1$$

and

$$(\Delta \otimes id)(J)(J \otimes 1) = (id \otimes \Delta)(J)(1 \otimes J).$$

We write

$$J = J^{(1)} \otimes J^{(2)} \quad \text{and} \quad J^{-1} = J^{-(1)} \otimes J^{-(2)},$$

where the summation is understood.

Given a twist J for H one can define a new Hopf algebra H^J with the same algebra structure and counit as H, for which the comultiplication Δ^J and antipode

S^J are given respectively by

$$\Delta^J(a) = J^{-1}\Delta(a)J,$$

$$S^J(a) = Q_J^{-1}S(a)Q_J, \quad \text{for } a \in H,$$

where $Q_J = S(J^{(1)})J^{(2)}$, which is an invertible element of H with the inverse $Q_J^{-1} = J^{-(1)}S(J^{-(2)})$. With the notations above, we have the following result:

Proposition 6.4.1 *Let H be a finite dimensional semisimple Hopf algebra over a field $\Bbbk$ of characteristic p with $\mathbf{u} = S(\Lambda_{(2)})\Lambda_{(1)}$ being invertible. The n-th Frobenius-Schur indicator $\nu_n(V)$ is invariant under twisting for any $n \in \mathbb{Z}$ and any finite dimensional H-module V.*

Proof Let Λ be a nonzero integral of H and J a normalized twist for H. It follows from [91, Theorem 3.4] that

$$P_n^J(\Lambda) = P_n(\Lambda),$$

where P_n^J and P_n are the n-th Sweedler power maps of H^J and H respectively. Moreover, $P_n(\Lambda)$ is a central element of H (see [91, Proposition 3.1]). Since

$$\Delta^J(\Lambda) = Q_J^{-1}\Lambda_{(1)} \otimes \Lambda_{(2)}Q_J,$$

it follows that

$$\mathbf{u}^J := S^J(\Lambda_{(2)}Q_J)Q_J^{-1}\Lambda_{(1)} = Q_J^{-1}S(Q_J)\mathbf{u}. \tag{6.14}$$

For H-module V with the character χ_V, we denote V^J the same as V as $\Bbbk$-linear space but thought of as an H^J-module. Then the character of V^J is also χ_V. For any $n \in \mathbb{Z}$, we have

$$\begin{aligned}
\nu_n(V^J) &= \chi_V((\mathbf{u}^J)^{-1}P_n^J(\Lambda)) \\
&= \chi_V(\mathbf{u}^{-1}S(Q_J^{-1})Q_J P_n^J(\Lambda)) \quad (\text{by } (6.14)) \\
&= \chi_V(\mathbf{u}^{-1}S(Q_J^{-1})Q_J P_n(\Lambda)) \\
&= \chi_V(\mathbf{u}^{-1}S^2(J^{-(2)})S(J^{-(1)})S(J^{(1)})J^{(2)}P_n(\Lambda)) \\
&= \chi_V(J^{-(2)}\mathbf{u}^{-1}S(J^{-(1)})S(J^{(1)})J^{(2)}P_n(\Lambda)) \\
&= \chi_V(\mathbf{u}^{-1}S(J^{-(1)})S(J^{(1)})J^{(2)}P_n(\Lambda)J^{-(2)}) \\
&= \chi_V(\mathbf{u}^{-1}S(J^{-(1)})S(J^{(1)})J^{(2)}J^{-(2)}P_n(\Lambda)) \\
&= \chi_V(\mathbf{u}^{-1}P_n(\Lambda)) \\
&= \nu_n(V).
\end{aligned}$$

This completes the proof. □

We are now ready to state the main result which says that higher Frobenius-Schur indicators are monoidal invariants of the representation category H-mod.

Theorem 6.4.2 *Let H and H' be two finite dimensional semisimple Hopf algebras over a field $\Bbbk$ of characteristic p with $\mathbf{u} = S(\Lambda_{(2)})\Lambda_{(1)}$ being invertible. If $\mathcal{F}$: H-mod $\to$ H'-mod is an equivalence of monoidal categories, then $\nu_n(V) = \nu_n(\mathcal{F}(V))$ for any $n \in \mathbb{Z}$ and any finite dimensional H-module V.*

Proof Since the $\Bbbk$-linear equivalence $\mathcal{F}$: H-mod $\to$ H'-mod is a monoidal equivalence, it follows from [63, Theorem 2.2] that there exists a twist J of H such that H' is isomorphic to H^J as bialgebras. Let $\sigma : H' \to H^J$ be such an isomorphism. Then σ is automatically a Hopf algebra isomorphism. The isomorphism σ induces a $\Bbbk$-linear equivalence $(-)^\sigma$: H-mod $\to$ H'-mod as follows: for any finite dimensional H-module V, $V^\sigma = V$ as $\Bbbk$-linear space with the H'-module action given by $h'v = \sigma(h')v$ for $h' \in H'$, $v \in V$, and $f^\sigma = f$ for any morphism f in H-mod. Moreover, the equivalence $\mathcal{F}$ is naturally isomorphic to the $\Bbbk$-linear equivalence $(-)^\sigma$ (see [42, Theorem 1.1]). Therefore,

$$\nu_n(\mathcal{F}(V)) = \nu_n(V^\sigma).$$

Let Λ' be a nonzero integral of H' such that $\sigma(\Lambda') = \Lambda$ and S' the antipode of H'. Note that the map $\sigma : H' \to H^J$ is a Hopf algebra isomorphism and Λ is a nonzero integral of H^J. It follows that

$$\sigma(P'_n(\Lambda')) = P^J_n(\Lambda),$$

where P'_n and P^J_n are the n-th Sweedler power maps of H' and H^J respectively. In particular,

$$\sigma((\mathbf{u}')^{-1}P'_n(\Lambda')) = (\mathbf{u}^J)^{-1}P^J_n(\Lambda),$$

where

$$\mathbf{u}' = S'(\Lambda'_{(2)})\Lambda'_{(1)} \quad \text{and} \quad \mathbf{u}^J = S^J(\Lambda_{\langle 2\rangle})\Lambda_{\langle 1\rangle}.$$

We have

$$\begin{aligned}
\nu_n(V^\sigma) &= \chi_{V^\sigma}((\mathbf{u}')^{-1}P'_n(\Lambda')) \\
&= \chi_{V^J}(\sigma((\mathbf{u}')^{-1}P'_n(\Lambda'))) \\
&= \chi_{V^J}((\mathbf{u}^J)^{-1}P^J_n(\Lambda)) \\
&= \nu_n(V^J) \\
&= \nu_n(V),
\end{aligned}$$

where the last equality follows from Proposition 6.4.1. We conclude that $\nu_n(\mathcal{F}(V)) = \nu_n(V)$ for any $n \in \mathbb{Z}$ and any finite dimensional H-module V. □

Chapter 7 The Grothendieck Algebras of Smash Product Hopf Algebras

Let H be a semisimple Hopf algebra over an algebraically closed field $\Bbbk$ of positive characteristic p. Under the conditions that $p > \dim_{\Bbbk}(H)^{1/2}$ and $p \nmid 2\dim_{\Bbbk}(H)$, we determine all non-isomorphic irreducible representations of the smash product semisimple Hopf algebra $H\#\Bbbk G$, where G is a cyclic group of order $n := 2\dim_{\Bbbk}(H)$. We endow with a new multiplication operator $\star$ on the Grothendieck algebra $(\mathrm{G}_{\Bbbk}(H), *)$ of H and show that the Grothendieck algebra $(\mathrm{G}_{\Bbbk}(H\#\Bbbk G), *)$ of $H\#\Bbbk G$ has the direct sum decomposition $(\mathrm{G}_{\Bbbk}(H), *)^{\oplus \frac{n}{2}} \bigoplus (\mathrm{G}_{\Bbbk}(H), \star)^{\oplus \frac{n}{2}}$. This reveals a relationship between the Grothendieck algebra of $H\#\Bbbk G$ and that of H.

7.1 Smash product Hopf algebras

Let H be a finite dimensional semisimple Hopf algebra over an algebraically closed field $\Bbbk$ of positive characteristic p with $p > \dim_{\Bbbk}(H)^{1/2}$ and $p \nmid 2\dim_{\Bbbk}(H)$. We denote by Λ and λ the left and right integrals of H and H^* respectively so that $\lambda(\Lambda) = 1$. Let G be a cyclic group of order $n = 2\dim_{\Bbbk}(H)$ generated by g.

Since the antipode S of H satisfies $S^{2n} = id$ by Radford's formula of S^4, the Hopf algebra H is a left $\Bbbk G$-module algebra whose action is given by

$$g^i \to h = S^{2i}(h), \quad \text{for } g^i \in G \text{ and } h \in H.$$

This reduces to a Hopf algebra $H\#\Bbbk G$ mentioned in [76]. More precisely, the Hopf algebra $H\#\Bbbk G$ is the smash product of H and $\Bbbk G$. The multiplication of $H\#\Bbbk G$ is given by

$$(a\#g^i)(b\#g^j) = a(g^i \to b)\#g^{i+j} = aS^{2i}(b)\#g^{i+j}, \quad \text{for } a, b \in H,$$

the identity of $H\#\Bbbk G$ is $1_H\#1_{\Bbbk G}$. The comultiplication of $H\#\Bbbk G$ is given by

$$\Delta_{H\#\Bbbk G}(h\#g^i) = (h_{(1)}\#g^i) \otimes (h_{(2)}\#g^i).$$

The counit of $H\#\Bbbk G$ is $\varepsilon_{H\#\Bbbk G} = \varepsilon_H\#\varepsilon_{\Bbbk G}$ and the antipode of $H\#\Bbbk G$ is given by

$$S_{H\#\Bbbk G}(h\#g^i) = (1_H\#g^{-i})(S(h)\#1_{\Bbbk G}) = S^{1-2i}(h)\#g^{-i}.$$

Moreover, $1_H\#g$ is a group-like element of $H\#\Bbbk G$ that satisfies

$$S^2_{H\#\Bbbk G}(h\#g^i) = (1_H\#g)(h\#g^i)(1_H\#g)^{-1}. \tag{7.1}$$

The Hopf algebra H can be considered as a sub-Hopf algebra of $H\#\Bbbk G$ under the injective map $H \to H\#\Bbbk G,\ h \mapsto h\#1_{\Bbbk G}$.

Since Λ is an integral of H with $\varepsilon(\Lambda) \neq 0$ and $p \nmid n$, $\Lambda\#\frac{1}{n}\sum_{i=0}^{n-1} g^i$ is an integral of $H\#\Bbbk G$ with

$$\varepsilon_{H\#\Bbbk G}(\Lambda\#\frac{1}{n}\sum_{i=0}^{n-1} g^i) = \varepsilon(\Lambda) \neq 0.$$

Thus, $H\#\Bbbk G$ is a semisimple Hopf algebra over $\Bbbk$.

We denote $\{V_i \mid 0 \leqslant i \leqslant m-1\}$ the set of all simple left H-modules up to isomorphism and $\{e_i \mid 0 \leqslant i \leqslant m-1\}$ the set of all central primitive idempotents of H. Note that V_0 is the trivial H-module $\Bbbk$ and e_0 is the idempotent $\Lambda/\varepsilon(\Lambda)$. The character of any simple H-module V_i is denoted by χ_i for $0 \leqslant i \leqslant m-1$ and the character of the left regular module H is denoted by χ_H. Obviously,

$$\chi_H = \sum_{i=0}^{m-1} \dim_\Bbbk(V_i)\chi_i.$$

Recall that the dual module V_i^* is also a simple H-module for $0 \leqslant i \leqslant m-1$. This induces a permutation $*$ on the index set $\{0, 1, \cdots, m-1\}$ defined by $i^* = j$ if $V_i^* \cong V_j$. The permutation $*$ satisfies that $i^{**} = i$, $S(e_i) = e_{i^*}$, $\dim_\Bbbk(V_{i^*}) = \dim_\Bbbk(V_i)$ and $\lambda(e_{i^*}) = \lambda(e_i)$ for $0 \leqslant i \leqslant m-1$ (see Remark 6.2.3).

We denote η_i to be a square root of $\lambda(e_i)/\varepsilon(\Lambda)$ for $0 \leqslant i \leqslant m-1$. Note that

$$\frac{1}{\varepsilon(\Lambda)^2} = \frac{\lambda(e_0)}{\varepsilon(\Lambda)} \quad \text{and} \quad \eta_{i^*}^2 = \frac{\lambda(e_{i^*})}{\varepsilon(\Lambda)} = \frac{\lambda(e_i)}{\varepsilon(\Lambda)} = \eta_i^2.$$

In view of this, we further assume that

- $\eta_0 = 1/\varepsilon(\Lambda)$;
- $\eta_i = \eta_{i^*}$ for $0 \leqslant i \leqslant m-1$.

We denote $\mathbf{u} := S(\Lambda_{(2)})\Lambda_{(1)}$ and

$$\mathbf{v} := \mathbf{u}\sum_{i=0}^{m-1} \frac{\eta_i}{\dim_\Bbbk(V_i)} e_i. \tag{7.2}$$

Here, $\dim_\Bbbk(V_i) \neq 0$ in $\Bbbk$ for $0 \leqslant i \leqslant m-1$ because of $p > \dim_\Bbbk(H)^{1/2}$. As we shall see, the element $\mathbf{v}$ plays a key role in the representation theory of the smash product Hopf algebra $H\#\Bbbk G$.

Proposition 7.1.1 *The element* $\mathbf{v}$ *satisfies the following properties:*

(1) $\varepsilon(\mathbf{v}) = 1$.

(2) $S^2(a) = \mathbf{v}a\mathbf{v}^{-1}$ *for* $a \in H$.

(3) $\mathbf{v}^2 = \mathbf{u}S(\mathbf{u}^{-1})$, *which is the distinguished group-like element* g_0 *of* H.

(4) $\mathbf{v}^n = 1$, *where* $n = 2\dim_{\mathbb{k}}(H)$.

(5) $\mathbf{v}^{-1} = S(\mathbf{v})$.

Proof (1) Note that $\eta_0 = 1/\varepsilon(\Lambda)$. Applying ε to both sides of the equality (7.2), we obtain that $\varepsilon(\mathbf{v}) = 1$.

(2) Since $S^2(a) = \mathbf{u}a\mathbf{u}^{-1}$ and the elements $\mathbf{u}$ and $\mathbf{v}$ are the same up to the central unit $\sum_{i=0}^{m-1} \frac{\eta_i}{\dim_{\mathbb{k}}(V_i)} e_i$, it follows that $S^2(a) = \mathbf{v}a\mathbf{v}^{-1}$ for $a \in H$.

(3) Note that

$$\mathbf{u}^{-1}S(\mathbf{u}^{-1}) = \frac{1}{\varepsilon(\Lambda)} \sum_{i=0}^{m-1} \frac{\lambda(e_i)}{\dim_{\mathbb{k}}(V_i)^2} e_i$$

by Proposition 6.2.1 (4). It follows that

$$\mathbf{u}S(\mathbf{u}^{-1}) = \frac{\mathbf{u}^2}{\varepsilon(\Lambda)} \sum_{i=0}^{m-1} \frac{\lambda(e_i)}{\dim_{\mathbb{k}}(V_i)^2} e_i = \mathbf{v}^2,$$

which is the distinguished group-like element g_0 of H by Proposition 6.2.1 (5).

(4) It can be seen from Part (3) that $\mathbf{v}^2$ is the distinguished group-like element g_0 of H, while the order of g_0 divides $\dim_{\mathbb{k}}(H)$. This implies that

$$\mathbf{v}^n = (\mathbf{v}^2)^{\dim_{\mathbb{k}}(H)} = 1.$$

(5) Note that

$$S(e_i) = e_{i^*}, \quad \dim_{\mathbb{k}}(V_{i^*}) = \dim_{\mathbb{k}}(V_i), \quad \eta_i = \eta_{i^*}, \quad \text{for } 0 \leqslant i \leqslant m-1.$$

We have

$$\begin{aligned}
\mathbf{v}S(\mathbf{v}) &= \mathbf{u}\left(\sum_{i=0}^{m-1} \frac{\eta_i}{\dim_{\mathbb{k}}(V_i)} e_i\right) S(\mathbf{u}) \left(\sum_{i=0}^{m-1} \frac{\eta_i}{\dim_{\mathbb{k}}(V_i)} e_{i^*}\right) \\
&= \mathbf{u}S(\mathbf{u}) \left(\sum_{i=0}^{m-1} \frac{\eta_i}{\dim_{\mathbb{k}}(V_i)} e_i\right) \left(\sum_{i=0}^{m-1} \frac{\eta_{i^*}}{\dim_{\mathbb{k}}(V_{i^*})} e_{i^*}\right) \\
&= \mathbf{u}S(\mathbf{u}) \left(\sum_{i=0}^{m-1} \frac{\eta_i}{\dim_{\mathbb{k}}(V_i)} e_i\right)^2 \\
&= \mathbf{u}S(\mathbf{u}) \left(\frac{1}{\varepsilon(\Lambda)} \sum_{i=0}^{m-1} \frac{\lambda(e_i)}{\dim_{\mathbb{k}}(V_i)^2} e_i\right)
\end{aligned}$$

$$= 1,$$

where the last equality follows from Proposition 6.2.1 (4). We obtain that $\mathbf{v}^{-1} = S(\mathbf{v})$. The proof is completed. □

7.2 Representations of smash product Hopf algebras

The representation theory of crossed product of an algebra with a group algebra has been studied in [57]. However, we do not take advantage of those notations and methods in [57] to describe $H\#\Bbbk G$-modules. Instead, since the Hopf algebra $H\#\Bbbk G$ is semisimple, we will determine all simple $H\#\Bbbk G$-modules by the study of the character of the regular representation of $H\#\Bbbk G$.

Lemma 7.2.1 *If V is a finite dimensional H-module and W is a finite dimensional $\Bbbk G$-module, then the vector space $V \otimes W$ is a finite dimensional $H\#\Bbbk G$-module, where the $H\#\Bbbk G$-module structure on $V \otimes W$ is given by*

$$(h\#g^k) \cdot (v \otimes w) = (h\mathbf{v}^k \cdot v) \otimes (g^k \cdot w), \quad \textit{for } v \in V, w \in W. \tag{7.3}$$

Proof By Proposition 7.1.1 (4), we have $\mathbf{v}^n = 1$. It follows that

$$(h\#g^n) \cdot (v \otimes w) = (h\mathbf{v}^n \cdot v) \otimes (g^n \cdot w) = (h \cdot v) \otimes w = (h\#1_{\Bbbk G}) \cdot (v \otimes w).$$

Hence the $H\#\Bbbk G$-module structure defined on $V \otimes W$ is compatible with the equality $h\#g^n = h\#1_{\Bbbk G}$. For $a, b \in H$, by $S^2(h) = \mathbf{v}h\mathbf{v}^{-1}$ for $h \in H$, we may check that

$$((a\#g^i)(b\#g^j)) \cdot (v \otimes w) = (a\#g^i) \cdot ((b\#g^j) \cdot (v \otimes w)).$$

Indeed,

$$\begin{aligned}((a\#g^i)(b\#g^j)) \cdot (v \otimes w) &= (aS^{2i}(b)\#g^{i+j}) \cdot (v \otimes w)\\ &= (aS^{2i}(b)\mathbf{v}^{i+j} \cdot v) \otimes (g^{i+j} \cdot w),\end{aligned}$$

while

$$\begin{aligned}(a\#g^i) \cdot ((b\#g^j) \cdot (v \otimes w)) &= (a\#g^i) \cdot ((b\mathbf{v}^j \cdot v) \otimes (g^j \cdot w))\\ &= (a\mathbf{v}^i \cdot (b\mathbf{v}^j \cdot v)) \otimes (g^i \cdot (g^j \cdot w))\\ &= (a\mathbf{v}^i b\mathbf{v}^j \cdot v) \otimes (g^{i+j} \cdot w)\\ &= (aS^{2i}(b)\mathbf{v}^{i+j} \cdot v) \otimes (g^{i+j} \cdot w).\end{aligned}$$

The proof is completed. □

Lemma 7.2.2 *If V is a simple H-module and W is a simple $\Bbbk G$-module, then $V \otimes W$ is a simple $H\#\Bbbk G$-module.*

Proof Note that $H\#\Bbbk G$ is a semisimple Hopf algebra over an algebraically closed field $\Bbbk$. It is sufficient to show that

$$\mathrm{End}_{H\#\Bbbk G}(V\otimes W)\cong \Bbbk.$$

Suppose that the map $\phi: V\otimes W\to V\otimes W$ is an $H\#\Bbbk G$-module morphism. Since W is one dimensional, we fix a basis w of W. The $H\#\Bbbk G$-module morphism ϕ induces an H-module morphism $\phi_0: V\to V$ as follows:

$$\phi(v\otimes w)=\phi_0(v)\otimes w,\quad \text{for any } v\in V.$$

This shows that ϕ is the identity map of $V\otimes W$ up to a scalar, since V is simple and ϕ_0 is the identity map of V up to a scalar. □

Note that the character group $\widehat{G}$ of G is a cyclic group of order n. Let ψ be a generator of $\widehat{G}$. Then $\widehat{G}=\{\psi^j \mid 0\leqslant j\leqslant n-1\}$, which is the complete set of distinct irreducible characters of simple $\Bbbk G$-modules. The simple $\Bbbk G$-module with respect to the character ψ^j is denoted by W_j for $0\leqslant j\leqslant n-1$.

Remark 7.2.3 *For simple H-module V_i and simple $\Bbbk G$-module W_j, it can be seen from Lemma 7.2.2 that $V_i\otimes W_j$ is a simple $H\#\Bbbk G$-module. Let χ_{ij} be the character associated to the simple $H\#\Bbbk G$-module $V_i\otimes W_j$. It follows from (7.3) that*

$$\chi_{ij}(h\otimes g^k)=\chi_i(h\mathbf{v}^k)\psi^j(g^k),\quad \textit{for } 0\leqslant i\leqslant m-1, 0\leqslant j\leqslant n-1.$$

Theorem 7.2.4 *The set $\{V_i\otimes W_j \mid 0\leqslant i\leqslant m-1, 0\leqslant j\leqslant n-1\}$ forms a complete set of non-isomorphic simple $H\#\Bbbk G$-modules.*

Proof Note that $\Lambda\#\frac{1}{n}\sum_{i=0}^{n-1}g^i$ is a left integral of $H\#\Bbbk G$ and $\lambda\#\sum_{j=0}^{n-1}\psi^j$ is a right integral of $(H\#\Bbbk G)^*$ satisfying

$$\left(\lambda\#\sum_{j=0}^{n-1}\psi^j\right)\left(\Lambda\#\frac{1}{n}\sum_{i=0}^{n-1}g^i\right)=1.$$

By (6.8), the characters of left regular representations of H and $H\#\Bbbk G$ are respectively given by

$$\chi_H=\lambda\leftharpoonup\mathbf{u}\quad\text{and}\quad \chi_{H\#\Bbbk G}=\left(\lambda\#\sum_{j=0}^{n-1}\psi^j\right)\leftharpoonup\mathbf{u}_{H\#\Bbbk G},$$

where

$$\mathbf{u}=S(\Lambda_{(2)})\Lambda_{(1)}$$

and

$$\mathbf{u}_{H\#\Bbbk G}=\frac{1}{n}\sum_{i=0}^{n-1}S_{H\#\Bbbk G}(\Lambda_{(2)}\#g^i)(\Lambda_{(1)}\#g^i)$$

$$
\begin{aligned}
&= \frac{1}{n}\sum_{i=0}^{n-1}(S^{1-2i}(\Lambda_{(2)})\#g^{-i})(\Lambda_{(1)}\#g^{i}) \\
&= \frac{1}{n}\sum_{i=0}^{n-1} S^{1-2i}(\Lambda_{(2)})S^{-2i}(\Lambda_{(1)})\#1_{\Bbbk G} \\
&= \frac{1}{n}\sum_{i=0}^{n-1} S^{-2i}(\mathbf{u})\#1_{\Bbbk G} \\
&= \mathbf{u}\#1_{\Bbbk G}.
\end{aligned}
$$

It follows that

$$
\chi_{H\#\Bbbk G} = \left(\lambda\#\sum_{j=0}^{n-1}\psi^j\right) \leftharpoonup (\mathbf{u}\#1_{\Bbbk G}) = (\lambda \leftharpoonup \mathbf{u})\#\sum_{j=0}^{n-1}\psi^j = \chi_H\#\sum_{j=0}^{n-1}\psi^j.
$$

Hence,

$$
(\chi_{H\#\Bbbk G})(h\#g^k) = \chi_H(h)\sum_{j=0}^{n-1}\psi^j(g^k) = \begin{cases} n\chi_H(h), & k=0, \\ 0, & 1 \leqslant k \leqslant n-1. \end{cases}
$$

While

$$
\begin{aligned}
\sum_{i=0}^{m-1}\sum_{j=0}^{n-1}\dim_{\Bbbk}(V_i\otimes W_j)\chi_{ij}(h\#g^k) &= \sum_{i=0}^{m-1}\sum_{j=0}^{n-1}\dim_{\Bbbk}(V_i)\chi_i(h\mathbf{v}^k)\psi^j(g^k) \\
&= \chi_H(h\mathbf{v}^k)\sum_{j=0}^{n-1}\psi^j(g^k) \\
&= \begin{cases} n\chi_H(h), & k=0, \\ 0, & 1 \leqslant k \leqslant n-1. \end{cases}
\end{aligned}
$$

We obtain that

$$
\chi_{H\#\Bbbk G} = \sum_{i=0}^{m-1}\sum_{j=0}^{n-1}\dim_{\Bbbk}(V_i\otimes W_j)\chi_{ij}.
$$

Hence, all non-isomorphic simple $H\#\Bbbk G$-modules are $V_i \otimes W_j$ for $0 \leqslant i \leqslant m-1$ and $0 \leqslant j \leqslant n-1$. □

Remark 7.2.5 *Note that $\chi_{00} = \varepsilon_{H\#\Bbbk G}$. Hence $V_0 \otimes W_0$ is the trivial $H\#\Bbbk G$-module, where V_0 is the trivial H-module and W_0 is the trivial $\Bbbk G$-module.*

For any simple $H\#\Bbbk G$-module $V_i \otimes W_j$, its dual module $(V_i \otimes W_j)^*$ can be described as follows:

Proposition 7.2.6 *We have*

$$(V_i \otimes W_j)^* \cong V_{i^*} \otimes W_{j^*}, \quad \text{for } 0 \leqslant i \leqslant m-1, 0 \leqslant j \leqslant n-1,$$

where V_{i^*} is the dual of V_i as an H-module and W_{j^*} is the dual of W_j as a $\Bbbk G$-module.

Proof We need to check that

$$\chi_{i^*j^*} = \chi_{ij} \circ S_{H\#\Bbbk G}, \quad \text{for } 0 \leqslant i \leqslant m-1, 0 \leqslant j \leqslant n-1.$$

Note that

$$S(\mathbf{v}) = \mathbf{v}^{-1} \quad \text{and} \quad S^{-2}(h) = \mathbf{v}^{-1} h \mathbf{v}, \quad \text{for } h \in H.$$

On the one hand,

$$\begin{aligned}
\chi_{i^*j^*}(h\#g^k) &= \chi_{i^*}(h\mathbf{v}^k)\psi^{j^*}(g^k) \\
&= \chi_i(S(\mathbf{v})^k S(h))\psi^j(g^{-k}) \\
&= \chi_i(\mathbf{v}^{-k} S(h))\psi^j(g^{-k}).
\end{aligned}$$

On the other hand,

$$\begin{aligned}
(\chi_{ij} \circ S_{H\#\Bbbk G})(h\#g^k) &= \chi_{ij}(S_{H\#\Bbbk G}(h\#g^k)) \\
&= \chi_{ij}(S^{1-2k}(h)\#g^{-k}) \\
&= \chi_{ij}(\mathbf{v}^{-k} S(h)\mathbf{v}^k \# g^{-k}) \\
&= \chi_i(\mathbf{v}^{-k} S(h))\psi^j(g^{-k}).
\end{aligned}$$

We conclude that $\chi_{i^*j^*} = \chi_{ij} \circ S_{H\#\Bbbk G}$ for $0 \leqslant i \leqslant m-1, 0 \leqslant j \leqslant n-1$. □

7.3 The Grothendieck algebras of smash product Hopf algebras

In this section, we will investigate a relationship between the Grothendieck algebra of the smash product Hopf algebra $H\#\Bbbk G$ and that of H. We still assume that the field $\Bbbk$ has positive characteristic p with $p > \dim_{\Bbbk}(H)^{1/2}$ and $p \nmid n$, where $n = 2\dim_{\Bbbk}(H)$.

Recall that the Grothendieck algebra $(G_{\Bbbk}(H), *)$ of H is an associative algebra over the field $\Bbbk$ with unity ε_H under the convolution $*$, where the convolution $*$ is defined by

$$(\chi_i * \chi_j)(h) = (\chi_i \otimes \chi_j)(\Delta(h)), \quad \text{for } h \in H.$$

We define a new multiplication operator $\star$ on $G_{\Bbbk}(H)$ by

$$(\chi_i \star \chi_j)(h) = (\chi_i \otimes \chi_j)\Big(\Delta(h)\Delta(\mathbf{v}^{-1})(\mathbf{v} \otimes \mathbf{v})\Big), \quad \text{for } h \in H.$$

Proposition 7.3.1 *The pair $(G_{\Bbbk}(H), \star)$ is an associative algebra over the field $\Bbbk$ with unity ε_H.*

Proof We first need to prove that $\star$ is a multiplication operator on $G_{\Bbbk}(H)$. That is, $\chi_i \star \chi_j \in G_{\Bbbk}(H)$ for $0 \leqslant i, j \leqslant m-1$. Indeed, for $a, b \in H$, using $S^2(h) = \mathbf{v}h\mathbf{v}^{-1}$ for $h \in H$, we have

$$
\begin{aligned}
(\chi_i \star \chi_j)(ab) &= \chi_i(a_{(1)}b_{(1)}\mathbf{v}^{-1}{}_{(1)}\mathbf{v})\chi_j(a_{(2)}b_{(2)}\mathbf{v}^{-1}{}_{(2)}\mathbf{v}) \\
&= \chi_i(a_{(1)}(b\mathbf{v}^{-1})_{(1)}\mathbf{v})\chi_j(a_{(2)}(b\mathbf{v}^{-1})_{(2)}\mathbf{v}) \\
&= \chi_i(a_{(1)}(\mathbf{v}^{-1}S^2(b))_{(1)}\mathbf{v})\chi_j(a_{(2)}(\mathbf{v}^{-1}S^2(b))_{(2)}\mathbf{v}) \\
&= \chi_i(a_{(1)}\mathbf{v}^{-1}{}_{(1)}S^2(b_{(1)})\mathbf{v})\chi_j(a_{(2)}\mathbf{v}^{-1}{}_{(2)}S^2(b_{(2)})\mathbf{v}) \\
&= \chi_i(a_{(1)}\mathbf{v}^{-1}{}_{(1)}\mathbf{v}b_{(1)})\chi_j(a_{(2)}\mathbf{v}^{-1}{}_{(2)}\mathbf{v}b_{(2)}) \\
&= \chi_i(b_{(1)}a_{(1)}\mathbf{v}^{-1}{}_{(1)}\mathbf{v})\chi_j(b_{(2)}a_{(2)}\mathbf{v}^{-1}{}_{(2)}\mathbf{v}) \\
&= (\chi_i \star \chi_j)(ba).
\end{aligned}
$$

It follows from [51] that

$$\chi_i \star \chi_j \in G_{\Bbbk}(H), \quad \text{for } 0 \leqslant i, j \leqslant m-1.$$

Since the map $H \to H \otimes H$, $h \mapsto \Delta(h)\Delta(\mathbf{v}^{-1})(\mathbf{v} \otimes \mathbf{v})$ is a coassociative comultiplication in H for which ε_H is still a counit (see [1, Eq.(12)]), the operator $\star$ dual to the coassociative comultiplication is an associative multiplication on $G_{\Bbbk}(H)$ with unity ε_H. The associativity and unity ε_H of $\star$ on $G_{\Bbbk}(H)$ can also be checked directly. Indeed, for $a \in H$, we have

$$
\begin{aligned}
((\chi_i \star \chi_j) \star \chi_k)(a) &= ((\chi_i \star \chi_j) \otimes \chi_k)\Big(\Delta(a)\Delta(\mathbf{v}^{-1})(\mathbf{v} \otimes \mathbf{v})\Big) \\
&= (\chi_i \star \chi_j)(a_{(1)}\mathbf{v}^{-1}{}_{(1)}\mathbf{v})\chi_k(a_{(2)}\mathbf{v}^{-1}{}_{(2)}\mathbf{v}) \\
&= \chi_i(a_{(1)}\mathbf{v}^{-1}{}_{(1)}\mathbf{v})\chi_j(a_{(2)}\mathbf{v}^{-1}{}_{(2)}\mathbf{v})\chi_k(a_{(3)}\mathbf{v}^{-1}{}_{(3)}\mathbf{v}) \\
&= \chi_i(a_{(1)}\mathbf{v}^{-1}{}_{(1)}\mathbf{v})(\chi_j \star \chi_k)(a_{(2)}\mathbf{v}^{-1}{}_{(2)}\mathbf{v}) \\
&= (\chi_i \star (\chi_j \star \chi_k))(a).
\end{aligned}
$$

Therefore,

$$(\chi_i \star \chi_j) \star \chi_k = \chi_i \star (\chi_j \star \chi_k), \quad \text{for } 0 \leqslant i, j, k \leqslant m-1.$$

$$
\begin{aligned}
(\varepsilon_H \star \chi_k)(a) &= (\varepsilon_H \otimes \chi_k)\Big(\Delta(a)\Delta(\mathbf{v}^{-1})(\mathbf{v} \otimes \mathbf{v})\Big) \\
&= \varepsilon_H(a_{(1)}\mathbf{v}^{-1}{}_{(1)}\mathbf{v})\chi_k(a_{(2)}\mathbf{v}^{-1}{}_{(2)}\mathbf{v})
\end{aligned}
$$

$$= \chi_k(a).$$

Hence

$$\varepsilon_H \star \chi_k = \chi_k, \quad \text{for } 0 \leqslant k \leqslant m-1.$$

It is similar that

$$\chi_k \star \varepsilon_H = \chi_k, \quad \text{for } 0 \leqslant k \leqslant m-1.$$

□

Next, we will use the algebras $(G_{\Bbbk}(H), *)$ and $(G_{\Bbbk}(H), \star)$ to describe the structure of the Grothendieck algebra $(G_{\Bbbk}(H\#\Bbbk G), *)$ of $H\#\Bbbk G$. Note that $\{\chi_0, \chi_1, \cdots, \chi_{m-1}\}$ is a $\Bbbk$-basis of $G_{\Bbbk}(H)$. Suppose in $(G_{\Bbbk}(H), *)$ that

$$\chi_i * \chi_j = \sum_{k=0}^{n-1} N_{ij}^k \chi_k$$

and in $(G_{\Bbbk}(H), \star)$ that

$$\chi_i \star \chi_j = \sum_{k=0}^{n-1} L_{ij}^k \chi_k,$$

where N_{ij}^k and L_{ij}^k are respectively the structure coefficients of the two algebras with respect to the basis $\{\chi_0, \chi_1, \cdots, \chi_{m-1}\}$. We stress that the coefficient N_{ij}^k is the multiplicity of V_k appeared in the decomposition of tensor product $V_i \otimes V_j$ as H-modules, so each N_{ij}^k is indeed a nonnegative integer. For the coefficient L_{ij}^k, we shall see in Remark 7.3.3 that each L_{ij}^k is an integer.

Proposition 7.3.2 *We have the following equations in the Grothendieck algebra* $(G_{\Bbbk}(H\#\Bbbk G), *)$:

(1) $\chi_{ij} = \chi_{i0} * \chi_{0j} = \chi_{0j} * \chi_{i0}$ *for* $0 \leqslant i \leqslant m-1, 0 \leqslant j \leqslant n-1$.

(2) $\chi_{i0} * \chi_{j0} = \sum_{k=0}^{m-1} \dfrac{1}{2}(N_{ij}^k + L_{ij}^k)\chi_{k0} + \sum_{k=0}^{m-1} \dfrac{1}{2}(N_{ij}^k - L_{ij}^k)\chi_{k\frac{n}{2}}$ *for* $0 \leqslant i, j \leqslant m-1$.

(3) $\chi_{is} * \chi_{jt} = \sum_{k=0}^{m-1} \dfrac{1}{2}(N_{ij}^k + L_{ij}^k)\chi_{k,s+t} + \sum_{k=0}^{m-1} \dfrac{1}{2}(N_{ij}^k - L_{ij}^k)\chi_{k,\frac{n}{2}+s+t}$ *for* $0 \leqslant i, j \leqslant m-1$ *and* $0 \leqslant s, t \leqslant n-1$, *where* $s+t$ *and* $\dfrac{n}{2}+s+t$ *are reduced modulo* n.

Proof (1) It is direct to calculate that

$$\begin{aligned}(\chi_{i0} * \chi_{0j})(h\#g^k) &= \chi_{i0}(h_{(1)}\#g^k)\chi_{0j}(h_{(2)}\#g^k) \\ &= \chi_i(h_{(1)}\mathbf{v}^k)\psi^0(g^k)\chi_0(h_{(2)}\mathbf{v}^k)\psi^j(g^k) \\ &= \chi_i(h\mathbf{v}^k)\psi^j(g^k) \\ &= \chi_{ij}(h\#g^k).\end{aligned}$$

So we have $\chi_{i0} * \chi_{0j} = \chi_{ij}$. It is similar that $\chi_{0j} * \chi_{i0} = \chi_{ij}$.

(2) We show that the values that both sides of the desired equation taking on $h\#g^l$ are the same. Note that $\mathbf{v}^2$ is the distinguished group-like element of H and $\psi^{\frac{n}{2}}(g) = -1$. For the case $l = 2s$, we have

$$\begin{aligned}
&\sum_{k=0}^{m-1} \frac{1}{2}(N_{ij}^k + L_{ij}^k)\chi_{k0}(h\#g^{2s}) + \sum_{k=0}^{m-1} \frac{1}{2}(N_{ij}^k - L_{ij}^k)\chi_{k\frac{n}{2}}(h\#g^{2s}) \\
&= \sum_{k=0}^{m-1} \frac{1}{2}(N_{ij}^k + L_{ij}^k)\chi_k(h\mathbf{v}^{2s}) + \sum_{k=0}^{m-1} \frac{1}{2}(N_{ij}^k - L_{ij}^k)\chi_k(h\mathbf{v}^{2s})\psi^{\frac{n}{2}}(g^{2s}) \\
&= \sum_{k=0}^{m-1} \frac{1}{2}(N_{ij}^k + L_{ij}^k)\chi_k(h\mathbf{v}^{2s}) + \sum_{k=0}^{m-1} \frac{1}{2}(N_{ij}^k - L_{ij}^k)\chi_k(h\mathbf{v}^{2s}) \\
&= \sum_{k=0}^{m-1} N_{ij}^k\chi_k(h\mathbf{v}^{2s}) \\
&= (\chi_i * \chi_j)(h\mathbf{v}^{2s}) \\
&= \chi_i(h_{(1)}\mathbf{v}^{2s})\chi_j(h_{(2)}\mathbf{v}^{2s}) \quad \text{(since } \mathbf{v}^{2s} \text{ is a group-like element)} \\
&= \chi_{i0}(h_{(1)}\#g^{2s})\chi_{j0}(h_{(2)}\#g^{2s}) \\
&= (\chi_{i0} * \chi_{j0})(h\#g^{2s}).
\end{aligned}$$

For the case $l = 2s + 1$, we have

$$\begin{aligned}
&\sum_{k=0}^{m-1} \frac{1}{2}(N_{ij}^k + L_{ij}^k)\chi_{k0}(h\#g^{2s+1}) + \sum_{k=0}^{m-1} \frac{1}{2}(N_{ij}^k - L_{ij}^k)\chi_{k\frac{n}{2}}(h\#g^{2s+1}) \\
&= \sum_{k=0}^{m-1} \frac{1}{2}(N_{ij}^k + L_{ij}^k)\chi_k(h\mathbf{v}^{2s+1}) + \sum_{k=0}^{m-1} \frac{1}{2}(N_{ij}^k - L_{ij}^k)\chi_k(h\mathbf{v}^{2s+1})\psi^{\frac{n}{2}}(g^{2s+1}) \\
&= \sum_{k=0}^{m-1} \frac{1}{2}(N_{ij}^k + L_{ij}^k)\chi_k(h\mathbf{v}^{2s+1}) - \sum_{k=0}^{m-1} \frac{1}{2}(N_{ij}^k - L_{ij}^k)\chi_k(h\mathbf{v}^{2s+1}) \\
&= \sum_{k=0}^{m-1} L_{ij}^k\chi_k(h\mathbf{v}^{2s+1}) \\
&= (\chi_i \star \chi_j)(h\mathbf{v}^{2s+1}) \\
&= \chi_i(h_{(1)}\mathbf{v}^{2s+1})\chi_j(h_{(2)}\mathbf{v}^{2s+1}) \quad \text{(since } \mathbf{v}^{2s} \text{ is a group-like element)} \\
&= \chi_{i0}(h_{(1)}\#g^{2s+1})\chi_{j0}(h_{(2)}\#g^{2s+1}) \\
&= (\chi_{i0} * \chi_{j0})(h\#g^{2s+1}).
\end{aligned}$$

We obtain the desired equation.

(3) Using Part (1) and Part (2) we may see that Part (3) holds. □

Remark 7.3.3 *It follows from Proposition* 7.3.2 (2) *that the tensor product* $(V_i \otimes W_0) \otimes (V_j \otimes W_0)$ *has the following decomposition as* $H\#\Bbbk G$*-modules*:

$$(V_i \otimes W_0) \otimes (V_j \otimes W_0) \cong \bigoplus_{k=0}^{m-1} \frac{1}{2}(N_{ij}^k + L_{ij}^k)(V_k \otimes W_0) \bigoplus \bigoplus_{k=0}^{m-1} \frac{1}{2}(N_{ij}^k - L_{ij}^k)(V_k \otimes W_{\frac{n}{2}}).$$

Thus, these coefficients $\frac{1}{2}(N_{ij}^k + L_{ij}^k)$ *and* $\frac{1}{2}(N_{ij}^k - L_{ij}^k)$ *are both nonnegative integers. Since all* N_{ij}^k *are nonnegative integers, it follows that all* L_{ij}^k *are integers and satisfy* $-N_{ij}^k \leqslant L_{ij}^k \leqslant N_{ij}^k$. *In view of this, the multiplication operator* $\star$ *defined on the Grothendieck algebra* $G_\Bbbk(H)$ *can be defined as well on the Grothendieck ring* $G_0(H)$.

The Grothendieck algebra $(G_\Bbbk(H\#\Bbbk G), *)$ is an associative unity algebra with a $\Bbbk$-basis $\{\chi_{ij} \mid 0 \leqslant i \leqslant m-1, 0 \leqslant j \leqslant n-1\}$. Denote by

$$\theta_l = \frac{1}{n}\sum_{t=0}^{n-1} \psi(g)^{-lt}\chi_{0t}, \quad \text{for } 0 \leqslant l \leqslant n-1.$$

Note that $\chi_{0t} = \psi^t$ for $0 \leqslant t \leqslant n-1$. Thus, $\{\theta_l \mid 0 \leqslant l \leqslant n-1\}$ is the set of all central primitive idempotents of the algebra $\Bbbk\widehat{G}$. Moreover, we have

$$\chi_{0j} * \theta_l = \psi(g)^{jl}\theta_l \quad \text{and} \quad \chi_{ij} * \theta_l = \chi_{i0} * \chi_{0j} * \theta_l = \psi(g)^{jl}\chi_{i0} * \theta_l. \tag{7.4}$$

In particular, each θ_l is a central idempotent of $(G_\Bbbk(H\#\Bbbk G), *)$. The structure of the Grothendieck algebra $(G_\Bbbk(H\#\Bbbk G), *)$ now can be described as follows:

Theorem 7.3.4 *We have the following algebra isomorphisms*:

(1) *If* l *is even, then* $(G_\Bbbk(H\#\Bbbk G), *) * \theta_l \cong (G_\Bbbk(H), *)$.

(2) *If* l *is odd, then* $(G_\Bbbk(H\#\Bbbk G), *) * \theta_l \cong (G_\Bbbk(H), \star)$.

(3) *We have* $(G_\Bbbk(H\#\Bbbk G), *) \cong (G_\Bbbk(H), *)^{\oplus \frac{n}{2}} \bigoplus (G_\Bbbk(H), \star)^{\oplus \frac{n}{2}}$.

Proof (1) For the case l being even, we consider the $\Bbbk$-linear map

$$\phi_l : (G_\Bbbk(H), *) \to (G_\Bbbk(H\#\Bbbk G), *) * \theta_l, \quad \chi_i \mapsto \chi_{i0} * \theta_l, \quad \text{for } \ 0 \leqslant i \leqslant m-1.$$

It can be seen from (7.4) that ϕ_l is bijective, and moreover, $\chi_{i\frac{n}{2}} * \theta_l = \chi_{i0} * \theta_l$. Now

$$\begin{aligned}\phi_l(\chi_i * \chi_j) &= \phi_l\left(\sum_{k=0}^{m-1} N_{ij}^k \chi_k\right)\\ &= \sum_{k=0}^{m-1} N_{ij}^k \chi_{k0} * \theta_l\end{aligned}$$

$$
\begin{aligned}
&= \sum_{k=0}^{m-1} \frac{1}{2}(N_{ij}^k + L_{ij}^k)\chi_{k0} * \theta_l + \sum_{k=0}^{m-1} \frac{1}{2}(N_{ij}^k - L_{ij}^k)\chi_{k0} * \theta_l \\
&= \sum_{k=0}^{m-1} \frac{1}{2}(N_{ij}^k + L_{ij}^k)\chi_{k0} * \theta_l + \sum_{k=0}^{m-1} \frac{1}{2}(N_{ij}^k - L_{ij}^k)\chi_{k\frac{n}{2}} * \theta_l \\
&= (\chi_{i0} * \chi_{j0}) * \theta_l \\
&= (\chi_{i0} * \theta_l) * (\chi_{j0} * \theta_l) \\
&= \phi_l(\chi_i) * \phi_l(\chi_j).
\end{aligned}
$$

This shows that ϕ_l is an algebra isomorphism.

(2) For the case l being odd, we consider the $\Bbbk$-linear map

$$\phi_l : (G_{\Bbbk}(H), \star) \to (G_{\Bbbk}(H\#\Bbbk G), *) * \theta_l, \quad \chi_i \mapsto \chi_{i0} * \theta_l, \quad \text{for } 0 \leqslant i \leqslant m-1.$$

It can be seen from (7.4) that ϕ_l is bijective, and moreover, $\chi_{i\frac{n}{2}} * \theta_l = -\chi_{i0} * \theta_l$. Now

$$
\begin{aligned}
\phi_l(\chi_i \star \chi_j) &= \phi_l\left(\sum_{k=0}^{m-1} L_{ij}^k \chi_k\right) \\
&= \sum_{k=0}^{m-1} L_{ij}^k \chi_{k0} * \theta_l \\
&= \sum_{k=0}^{m-1} \frac{1}{2}(N_{ij}^k + L_{ij}^k)\chi_{k0} * \theta_l - \sum_{k=0}^{m-1} \frac{1}{2}(N_{ij}^k - L_{ij}^k)\chi_{k0} * \theta_l \\
&= \sum_{k=0}^{m-1} \frac{1}{2}(N_{ij}^k + L_{ij}^k)\chi_{k0} * \theta_l + \sum_{k=0}^{m-1} \frac{1}{2}(N_{ij}^k - L_{ij}^k)\chi_{k\frac{n}{2}} * \theta_l \\
&= (\chi_{i0} * \chi_{j0}) * \theta_l \\
&= (\chi_{i0} * \theta_l) * (\chi_{j0} * \theta_l) \\
&= \phi_l(\chi_i) * \phi_l(\chi_j).
\end{aligned}
$$

Thus, ϕ_l is an algebra isomorphism.

(3) Let $(G_{\Bbbk}(H), *)^{\oplus\frac{n}{2}}$ be the direct sum of $\frac{n}{2}$-folds of $(G_{\Bbbk}(H), *)$ and $(G_{\Bbbk}(H), \star)^{\oplus\frac{n}{2}}$ the direct sum of $\frac{n}{2}$-folds of $(G_{\Bbbk}(H), \star)$. Since $\theta_0 + \theta_1 + \cdots + \theta_{n-1} = 1$, where 1 is the unity χ_{00} of $(G_{\Bbbk}(H\#\Bbbk G), *)$, using Part (1) and Part (2) we obtain the following algebra isomorphism:

$$(G_{\Bbbk}(H\#\Bbbk G), *) \cong (G_{\Bbbk}(H), *)^{\oplus\frac{n}{2}} \bigoplus (G_{\Bbbk}(H), \star)^{\oplus\frac{n}{2}}.$$

The proof is completed. □

Remark 7.3.5 *If $S^2 = id$, then $\mathbf{u} = \varepsilon(\Lambda)$ and $\lambda(e_i)/\varepsilon(\Lambda) = (\dim_{\Bbbk}(V_i)/\varepsilon(\Lambda))^2$ by Proposition 6.2.1 (3). Now η_i, as a square root of $\lambda(e_i)/\varepsilon(\Lambda)$, may be chosen to be $\dim_{\Bbbk}(V_i)/\varepsilon(\Lambda)$. It follows that*

$$\mathbf{v} = \mathbf{u}\sum_{i=0}^{m-1} \frac{\eta_i}{\dim_{\Bbbk}(V_i)} e_i = 1.$$

In this case, The multiplication operator $\star$ considered above is nothing but the convolution $*$ and the algebra $(G_{\Bbbk}(H), \star)$ is nothing but the Grothendieck algebra $(G_{\Bbbk}(H), *)$. Moreover,

$$(G_{\Bbbk}(H\#\Bbbk G), *) \cong (G_{\Bbbk}(H), *)^{\oplus\frac{n}{2}} \bigoplus (G_{\Bbbk}(H), \star)^{\oplus\frac{n}{2}} \cong (G_{\Bbbk}(H), *)^{\oplus n}.$$

Let $\mathcal{C}$ be the $\Bbbk$-linear subcategory of $H\#\Bbbk G$-mod spanned by objects

$$\{V_i \otimes W_0, V_i \otimes W_{\frac{n}{2}} \mid 0 \leqslant i \leqslant m-1\}.$$

Then $\mathcal{C}$ is closed under taking dual by Proposition 7.2.6. It follows from Proposition 7.3.2 that $\mathcal{C}$ is also closed under the tensor product of objects. More explicitly,

$$\begin{aligned}(V_i \otimes W_{\frac{n}{2}}) \otimes (V_j \otimes W_{\frac{n}{2}}) &\cong (V_i \otimes W_0) \otimes (V_j \otimes W_0)\\ &\cong \bigoplus_{k=0}^{m-1} \frac{1}{2}(N_{ij}^k + L_{ij}^k)(V_k \otimes W_0) \bigoplus \bigoplus_{k=0}^{m-1} \frac{1}{2}(N_{ij}^k - L_{ij}^k)(V_k \otimes W_{\frac{n}{2}}),\end{aligned}$$

and

$$\begin{aligned}(V_i \otimes W_0) \otimes (V_j \otimes W_{\frac{n}{2}}) &\cong (V_i \otimes W_{\frac{n}{2}}) \otimes (V_j \otimes W_0)\\ &\cong \bigoplus_{k=0}^{m-1} \frac{1}{2}(N_{ij}^k + L_{ij}^k)(V_k \otimes W_{\frac{n}{2}}) \bigoplus \bigoplus_{k=0}^{m-1} \frac{1}{2}(N_{ij}^k - L_{ij}^k)(V_k \otimes W_0).\end{aligned}$$

Hence $\mathcal{C}$ is a fusion subcategory of $H\#\Bbbk G$-mod. Let $(G_{\Bbbk}(\mathcal{C}), *)$ be the Grothendieck algebra of $\mathcal{C}$. Then $\{\chi_{i0}, \chi_{i\frac{n}{2}} \mid 0 \leqslant i \leqslant m-1\}$ forms a $\Bbbk$-basis of $(G_{\Bbbk}(\mathcal{C}), *)$.

Proposition 7.3.6 *We have the following algebra isomorphism:*

$$(G_{\Bbbk}(\mathcal{C}), *) \cong (G_{\Bbbk}(H), *) \bigoplus (G_{\Bbbk}(H), \star).$$

Proof We denote $\theta = \frac{1}{2}(\chi_{00} + \chi_{0\frac{n}{2}})$. Then $1 - \theta = \frac{1}{2}(\chi_{00} - \chi_{0\frac{n}{2}})$, where 1 is the unity χ_{00} of $(G_{\Bbbk}(\mathcal{C}), *)$. Note that θ and $1-\theta$ are both central idempotents of $(G_{\Bbbk}(\mathcal{C}), *)$. In particular,

$$\chi_{i\frac{n}{2}} * \theta = \chi_{i0} * \chi_{0\frac{n}{2}} * \theta = \chi_{i0} * \theta, \quad \text{for } 0 \leqslant i \leqslant m-1.$$

Consider the $\Bbbk$-linear map

$$\phi : (G_{\Bbbk}(H), *) \to (G_{\Bbbk}(\mathcal{C}), *) * \theta, \quad \chi_i \mapsto \chi_{i0} * \theta, \quad \text{for } 0 \leqslant i \leqslant m-1.$$

It is easy to see that ϕ is bijective and

$$\begin{aligned}
\phi(\chi_i * \chi_j) &= \phi\left(\sum_{k=0}^{m-1} N_{ij}^k \chi_k\right) \\
&= \sum_{k=0}^{m-1} N_{ij}^k \chi_{k0} * \theta \\
&= \sum_{k=0}^{m-1} \frac{1}{2}(N_{ij}^k + L_{ij}^k)\chi_{k0} * \theta + \sum_{k=0}^{m-1} \frac{1}{2}(N_{ij}^k - L_{ij}^k)\chi_{k0} * \theta \\
&= \sum_{k=0}^{m-1} \frac{1}{2}(N_{ij}^k + L_{ij}^k)\chi_{k0} * \theta + \sum_{k=0}^{m-1} \frac{1}{2}(N_{ij}^k - L_{ij}^k)\chi_{k\frac{n}{2}} * \theta \\
&= (\chi_{i0} * \chi_{j0}) * \theta \\
&= (\chi_{i0} * \theta) * (\chi_{j0} * \theta) \\
&= \phi(\chi_i) * \phi(\chi_j).
\end{aligned}$$

This shows that ϕ is an algebra isomorphism. Consider the $\Bbbk$-linear map

$$\varphi : (G_{\Bbbk}(H), \star) \to (G_{\Bbbk}(\mathcal{C}), *) * (1-\theta), \quad \chi_i \mapsto \chi_{i0} * (1-\theta), \quad \text{for } 0 \leqslant i \leqslant m-1.$$

Then φ is bijective. Using $\chi_{i\frac{n}{2}} * (1-\theta) = -\chi_{i0} * (1-\theta)$ we may see that

$$\begin{aligned}
\varphi(\chi_i \star \chi_j) &= \varphi\left(\sum_{k=0}^{m-1} L_{ij}^k \chi_k\right) \\
&= \sum_{k=0}^{m-1} L_{ij}^k \chi_{k0} * (1-\theta) \\
&= \sum_{k=0}^{m-1} \frac{1}{2}(N_{ij}^k + L_{ij}^k)\chi_{k0} * (1-\theta) - \sum_{k=0}^{m-1} \frac{1}{2}(N_{ij}^k - L_{ij}^k)\chi_{k0} * (1-\theta) \\
&= \sum_{k=0}^{m-1} \frac{1}{2}(N_{ij}^k + L_{ij}^k)\chi_{k0} * (1-\theta) + \sum_{k=0}^{m-1} \frac{1}{2}(N_{ij}^k - L_{ij}^k)\chi_{k\frac{n}{2}} * (1-\theta) \\
&= (\chi_{i0} * \chi_{j0}) * (1-\theta) \\
&= (\chi_{i0} * (1-\theta)) * (\chi_{j0} * (1-\theta))
\end{aligned}$$

$$= \varphi(\chi_i) * \varphi(\chi_j).$$

Hence, φ is an algebra isomorphism. □

Note that

$$\theta = \theta_0 + \theta_2 + \theta_4 + \cdots + \theta_{n-2} \text{ and } 1 - \theta = \theta_1 + \theta_3 + \theta_5 + \cdots + \theta_{n-1}.$$

By Theorem 7.3.4 and Proposition 7.3.6, we have the following corollary:

Corollary 7.3.7 *We have the following algebra isomorphism:*

$$(G_{\Bbbk}(H\#\Bbbk G), *) \cong (G_{\Bbbk}(\mathcal{C}), *)^{\oplus \frac{n}{2}}.$$

Finally, we give some remarks on the pivotal (spherical) structures of the fusion categories $H\#\Bbbk G$-mod and $\mathcal{C}$. Since $S^2_{H\#\Bbbk G}$ is an inner automorphism of $H\#\Bbbk G$ and

$$S^2_{H\#\Bbbk G}(h\#g^i) = (1_H\#g)(h\#g^i)(1_H\#g)^{-1},$$

where $1_H\#g$ is a group-like element of $H\#\Bbbk G$, the representation category $H\#\Bbbk G$-mod is a pivotal fusion category, where the pivotal structure τ on $H\#\Bbbk G$-mod is the isomorphism of monoidal functors $\tau_{V\otimes W} : V \otimes W \to (V\otimes W)^{**}$ natural in $V\otimes W$. Here $\tau_{V\otimes W}(v\otimes w)$ is defined by

$$\tau_{V\otimes W}(v\otimes w)(f) = f(1_H\#g \cdot v \otimes w) = f(\mathbf{v}\cdot v \otimes g \cdot w),$$

for $v \in V, w \in W$ and $f \in (V\otimes W)^*$.

The quantum dimension of the representation $V\otimes W$ of $H\#\Bbbk G$ with respect to the pivotal structure τ is denoted by $\mathbf{dim}(V\otimes W)$, which is the following composition

$$\mathbf{1} \xrightarrow{\mathrm{coev}_{(V\otimes W)}} (V\otimes W)\otimes(V\otimes W)^* \xrightarrow{\tau_{V\otimes W}\otimes id} (V\otimes W)^{**}\otimes(V\otimes W)^* \xrightarrow{\mathrm{ev}_{(V\otimes W)^*}} \mathbf{1},$$

where $\mathbf{1}$ is the trivial $H\#\Bbbk G$-module $V_0\otimes W_0$. From this composition, we have

$$\mathbf{dim}(V\otimes W) = \chi_V(\mathbf{v})\chi_W(g).$$

Especially,

$$\mathbf{dim}(V_i\otimes W_j) = \chi_i(\mathbf{v})\psi^j(g) = \varepsilon(\Lambda)\eta_i\psi^j(g).$$

For the dual module

$$(V_i\otimes W_j)^* \cong V_{i^*}\otimes W_{j^*},$$

we have

$$\mathbf{dim}(V_{i^*}\otimes W_{j^*}) = \varepsilon(\Lambda)\eta_{i^*}\psi^j(g^{-1}) = \varepsilon(\Lambda)\eta_i\psi^j(g^{-1}).$$

Therefore,

$$\mathbf{dim}(V_{i^*}\otimes W_{j^*}) = \mathbf{dim}(V_i\otimes W_j)$$

if and only if $\psi^j(g) = \psi^j(g^{-1})$, if and only if$j = 0$ or $j = \frac{n}{2}$. This means that with respect to the pivotal structure τ, the fusion category $H\#\Bbbk G$-mod is pivotal but not spherical, while the the fusion subcategory $\mathcal{C}$ of $H\#\Bbbk G$-mod spanned by objects $\{V_i \otimes W_0, V_i \otimes W_{\frac{n}{2}} \mid 0 \leqslant i \leqslant m-1\}$ is both pivotal and spherical.

Chapter 8 Invariants from the Sweedler Power Maps on Integrals

Throughout this chapter, H is an arbitrary finite dimensional Hopf algebra over a ground field $\Bbbk$. The inverse of the antipode S of H under composition is denoted by S^{-1}. We fix a left integral $\Lambda \in H$ and a right integral $\lambda \in H^*$ such that $\lambda(\Lambda) = 1$. We denote α to be the distinguished group-like element of H^* defined by $\Lambda a = \alpha(a)\Lambda$ for $a \in H$. The category of finite dimensional left H-modules is denoted by H-mod. In this chapter, we concern with those polynomials vanishing on certain images of the Sweedler power maps on the integral Λ. We show that these polynomials are monoidal invariants of H-mod. As an application, we show that the representation categories of 12-dimensional pointed non-semisimple Hopf algebras are mutually inequivalent as monoidal categories. Also, we distinguish the representation categories K_8-mod, $\Bbbk Q_8$-mod and $\Bbbk D_4$-mod whereas they have the same fusion rules. We finally use results obtained in this chapter to give a uniform proof of the remarkable result which says that the n-th indicator of a Hopf algebra is a monoidal invariant for any $n \in \mathbb{Z}$.

8.1 The Sweedler power maps on integrals

In this section, we first give a property of $P_n(\Lambda)$ for a left integral $\Lambda \in H$. We then use this property to describe a relationship between $P_n^J(\Lambda)$ and $P_n(\Lambda)$ for a normalized twist J of H.

The following equalities are of fundamental importance. The first equality can be found in [46, Lemma 1.2]:

$$S(a)\Lambda_{(1)} \otimes \Lambda_{(2)} = \Lambda_{(1)} \otimes a\Lambda_{(2)}, \quad \text{for } a \in H. \tag{8.1}$$

The second can be found in [42]:

$$\Lambda_{(1)}a \otimes \Lambda_{(2)} = \Lambda_{(1)} \otimes \Lambda_{(2)}\alpha(a_{(1)})S(a_{(2)}), \quad \text{for } a \in H, \tag{8.2}$$

where α is the distinguished group-like element of H^*. For $a, b \in H$, the integral λ satisfies the following properties:

$$\lambda(ab) = \lambda(\alpha(b_{(1)})S^2(b_{(2)})a), \tag{8.3}$$

$$\lambda(ab_{(1)})b_{(2)} = \lambda(a_{(1)}b)S^{-1}(a_{(2)}), \tag{8.4}$$

where (8.3) follows from [70, Theorem 3(a)] and (8.4) follows from [1, Remark 3.2]. For any $\Bbbk$-linear map $f : H \to H$, the trace of f can be described by Radford's trace formula (see [70, Theorem 2]):

$$\operatorname{tr}(f) = \lambda(S(\Lambda_{(2)})f(\Lambda_{(1)})). \tag{8.5}$$

Recall from [45] that the n-th Sweedler power map $P_n : H \to H$ is the n-th convolution power of the identity map id of H. Namely,

$$P_n(a) = \begin{cases} a_{(1)} \cdots a_{(n)}, & n \geqslant 1, \\ \varepsilon(a), & n = 0, \\ S(a_{(-n)} \cdots a_{(1)}), & n \leqslant -1. \end{cases}$$

Note that $P_1(a) = a$ and $P_{-1}(a) = S(a)$ for $a \in H$.

Recall that a normalized twist for a finite dimensional Hopf algebra H is an invertible element $J \in H \otimes H$ that satisfies

$$(\varepsilon \otimes id)(J) = (id \otimes \varepsilon)(J) = 1$$

and

$$(\Delta \otimes id)(J)(J \otimes 1) = (id \otimes \Delta)(J)(1 \otimes J). \tag{8.6}$$

We write

$$J = J^{(1)} \otimes J^{(2)} \quad \text{and} \quad J^{-1} = J^{-(1)} \otimes J^{-(2)},$$

where the summation is understood. We also write

$$J_{21} = J^{(2)} \otimes J^{(1)}.$$

Given a normalized twist J for H one can define a new Hopf algebra H^J with the same algebra structure and counit as H, for which the comultiplication Δ^J and antipode S^J are given respectively by

$$\Delta^J(a) = J^{-1}\Delta(a)J, \qquad S^J(a) = Q_J^{-1}S(a)Q_J, \quad \text{for } a \in H,$$

where $Q_J = S(J^{(1)})J^{(2)}$, which is an invertible element of H with the inverse $Q_J^{-1} = J^{-(1)}S(J^{-(2)})$.

The element Q_J satisfies the following identity (see [1, Eq (5)]):

$$\Delta(Q_J) = (S \otimes S)(J_{21}^{-1})(Q_J \otimes Q_J)J^{-1}. \tag{8.7}$$

This implies that

$$\Delta(Q_J^{-1}) = J(Q_J^{-1} \otimes Q_J^{-1})(S \otimes S)(J_{21}). \tag{8.8}$$

The twist J of H satisfies the following properties (see [1, Lemma 2.4]):

$$S(J^{(1)}_{(1)}) \otimes S(J^{(1)}_{(2)})J^{(2)} = (S \otimes S)(J^{-1})(1 \otimes Q_J), \tag{8.9}$$

$$J^{-(1)}_{(1)} \otimes J^{-(1)}_{(2)} S(J^{-(2)}) = J(1 \otimes Q_J^{-1}), \tag{8.10}$$

$$(Q_J^{-1} \otimes 1)(S \otimes S)(J) = J^{-(1)} S(J^{-(2)}_{(1)}) \otimes S(J^{-(2)}_{(2)}). \tag{8.11}$$

Applying $S^{-1} \otimes id$ to both sides of the equality (8.9), we have

$$J^{(1)}_{(1)} \otimes S(J^{(1)}_{(2)})J^{(2)} = J^{-(1)} \otimes S(J^{-(2)})Q_J. \tag{8.12}$$

Taking the ordinary flip map on both sides of the equality (8.11), we have

$$(1 \otimes Q_J^{-1})(S \otimes S)(J_{21}) = (1 \otimes J^{-(1)})\Delta(S(J^{-(2)})). \tag{8.13}$$

We define

$$\Delta^{(1)} = id \quad \text{and} \quad \Delta^{(n+1)} = (id \otimes \Delta^{(n)}) \circ \Delta, \quad \text{for all } n \geqslant 1.$$

We denote

$$\Delta^{(n)}(a) = a_{(1)} \otimes \cdots \otimes a_{(n)} \quad \text{and} \quad (\Delta^J)^{(n)}(a) = a_{\langle 1\rangle} \otimes \cdots \otimes a_{\langle n\rangle}$$

to distinguish between $\Delta^{(n)}(a)$ and $(\Delta^J)^{(n)}(a)$ for $a \in H$. The n-th Sweedler power map of H^J is denoted by P_n^J. Then $P_n^J(a)$ can be written as

$$P_n^J(a) = \begin{cases} a_{\langle 1\rangle} \cdots a_{\langle n\rangle}, & n \geqslant 1, \\ \varepsilon(a), & n = 0, \\ S^J(a_{\langle -n\rangle} \cdots a_{\langle 1\rangle}), & n \leqslant -1. \end{cases}$$

Proposition 8.1.1 *Let H be a finite dimensional Hopf algebra over a field $\Bbbk$ with a nonzero left integral Λ. For any $n \in \mathbb{Z}$ and $a \in H$, we have*

$$aP_n(\Lambda) = P_n(\Lambda)a^\dagger,$$

where $a^\dagger = a_{(1)}(\alpha \circ S^{-1})(a_{(2)})$. Moreover, if H is unimodular, then $P_n(\Lambda)$ is a central element of H.

Proof Obviously, the desired result holds for the cases $n = 0, \pm 1$. For the case $n > 1$, we have

$$\begin{aligned} aP_n(\Lambda) &= a\Lambda_{(1)}P_{n-1}(\Lambda_{(2)}) \\ &= \Lambda_{(1)}P_{n-1}(S^{-1}(a)\Lambda_{(2)}) \ \text{(by (8.1))} \end{aligned}$$

$$
\begin{aligned}
&= \Lambda_{(1)}(S^{-1}(a))_{(1)}\Lambda_{(2)}(S^{-1}(a))_{(2)}\cdots\Lambda_{(n-1)}(S^{-1}(a))_{(n-1)}\Lambda_{(n)} \\
&= P_{n-1}(\Lambda_{(1)}S^{-1}(a))\Lambda_{(2)} \\
&= P_{n-1}(\Lambda_{(1)})\Lambda_{(2)}a_{(1)}\alpha(S^{-1}(a_{(2)})) \quad \text{(by (8.2))} \\
&= P_n(\Lambda)a^{\dagger}.
\end{aligned}
$$

For the case $n < -1$, we have

$$
\begin{aligned}
aP_n(\Lambda) &= S(\Lambda_{(1)}S^{-1}(a))P_{n+1}(\Lambda_{(2)}) \\
&= S(\Lambda_{(1)})P_{n+1}(\Lambda_{(2)}a^{\dagger}) \quad \text{(by (8.2))} \\
&= S(\Lambda_{(1)})S(\Lambda_{(-n)}a^{\dagger}_{(-n-1)}\cdots\Lambda_{(2)}a^{\dagger}_{(1)}) \\
&= S(a^{\dagger}_{(-n-1)}\Lambda_{(-n-1)}\cdots a^{\dagger}_{(1)}\Lambda_{(1)})S(\Lambda_{(-n)}) \\
&= P_{n+1}(a^{\dagger}\Lambda_{(1)})S(\Lambda_{(2)}) \\
&= P_{n+1}(\Lambda_{(1)})S(S^{-1}(a^{\dagger})\Lambda_{(2)}) \qquad \text{(by (8.1))} \\
&= S(\Lambda_{(-n-1)}\cdots\Lambda_{(1)})S(\Lambda_{(-n)})a^{\dagger} \\
&= P_n(\Lambda)a^{\dagger}.
\end{aligned}
$$

If H is unimodular, then $\alpha = \varepsilon$ and hence $a^{\dagger} = a$. In this case, $P_n(\Lambda)$ is a central element of H. □

Remark 8.1.2 *For any $a \in H$, we define a family of elements $a_k \in H$ recursively by*

$$
a_1 = a_{(1)}(\alpha\circ S^{-1})(a_{(2)}) \quad \textit{and} \quad a_{k+1} = (a_k)_{(1)}(\alpha\circ S^{-1})((a_k)_{(2)}), \quad \textit{for } k \geqslant 1. \tag{8.14}
$$

By induction on k, we may see that

$$
a(P_n(\Lambda))^k = (P_n(\Lambda))^k a_k, \quad \textit{for any } n \in \mathbb{Z}.
$$

This further implies that

$$
(aP_n(\Lambda))^k = (P_n(\Lambda))^k a_k \cdots a_2 a_1.
$$

Moreover, for positive integers $k_1, \cdots, k_s$ and integers $n_1, \cdots, n_s$, we have

$$
(aP_{n_1}(\Lambda))^{k_1}\cdots(aP_{n_s}(\Lambda))^{k_s} = (P_{n_1}(\Lambda))^{k_1}\cdots(P_{n_s}(\Lambda))^{k_s}a_{k_1+\cdots+k_s}\cdots a_2 a_1. \tag{8.15}
$$

To describe a relationship between $P_n^J(\Lambda)$ and $P_n(\Lambda)$, we need the following lemma:

Lemma 8.1.3 *Let H be a finite dimensional Hopf algebra over a field $\Bbbk$ with a normalized twist J and a nonzero left integral Λ. Denote by $R=\alpha(J^{-(1)})J^{-(2)}$ and $T=J^{-(1)}\alpha(J^{-(2)})$. For any integer $n\geqslant 2$, we have*

(1)

$$\begin{aligned}&\Lambda_{\langle 1\rangle}\otimes\Lambda_{\langle 2\rangle}\Lambda_{\langle 3\rangle}\cdots\Lambda_{\langle n\rangle}\\=&Q_J^{-1}\Lambda_{(1)}J^{(1)}\otimes J^{-(1)}\Lambda_{(2)}P_{n-2}(J^{(2)}J^{-(2)}\Lambda_{(3)})S(R)Q_J.\end{aligned}$$

(2)

$$\begin{aligned}&\Lambda_{\langle n\rangle}\otimes\Lambda_{\langle n-1\rangle}\Lambda_{\langle n-2\rangle}\cdots\Lambda_{\langle 1\rangle}\\=&S^{-1}(Q_J^{-1})\Lambda_{(3)}J^{(2)}\otimes J^{-(2)}\Lambda_{(2)}(S^{-1}\circ P_{2-n})(J^{(1)}J^{-(1)}\Lambda_{(1)})S^{-1}(T)S^{-1}(Q_J).\end{aligned}$$

Proof We only give a proof of Part (1) and the proof of Part (2) is similar to that of Part (1). We proceed by induction on n. For the case $n=2$,

$$\begin{aligned}\Lambda_{\langle 1\rangle}\otimes\Lambda_{\langle 2\rangle}&=\Delta^J(\Lambda)=J^{-1}\Delta(\Lambda)J=J^{-(1)}\Lambda_{(1)}J^{(1)}\otimes J^{-(2)}\Lambda_{(2)}J^{(2)}\\&=J^{-(1)}S(J^{-(2)})\Lambda_{(1)}J^{(1)}\otimes\Lambda_{(2)}J^{(2)}\quad\text{(by (8.1))}\\&=J^{-(1)}S(J^{-(2)})\Lambda_{(1)}\otimes\Lambda_{(2)}\alpha(J^{(1)}_{(1)})S(J^{(1)}_{(2)})J^{(2)}\quad\text{(by (8.2))}\\&=Q_J^{-1}\Lambda_{(1)}\otimes\Lambda_{(2)}S(R)Q_J\quad\text{(by (8.12))}.\end{aligned}$$

So the equality holds for the case $n=2$. Suppose the identity in question holds for the case n. Namely,

$$\begin{aligned}&\Lambda_{\langle 1\rangle}\otimes\Lambda_{\langle 2\rangle}\Lambda_{\langle 3\rangle}\cdots\Lambda_{\langle n\rangle}\\=&Q_J^{-1}\Lambda_{(1)}J^{(1)}\otimes J^{-(1)}\Lambda_{(2)}P_{n-2}(J^{(2)}J^{-(2)}\Lambda_{(3)})S(R)Q_J.\end{aligned}\tag{8.16}$$

To prove the case $n+1$, we set $\widetilde{J}=J$. Namely,

$$\widetilde{J}^{(1)}\otimes\widetilde{J}^{(2)}=J^{(1)}\otimes J^{(2)},$$

or equivalently

$$\widetilde{J}^{-(1)}\otimes\widetilde{J}^{-(2)}=J^{-(1)}\otimes J^{-(2)}.$$

Applying $\Delta^J\otimes id$ to both sides of (8.16), we have

$$\begin{aligned}&\Lambda_{\langle 1\rangle}\otimes\Lambda_{\langle 2\rangle}\otimes\Lambda_{\langle 3\rangle}\cdots\Lambda_{\langle n+1\rangle}\\&=J^{-1}\Delta(Q_J^{-1})\Delta(\Lambda_{(1)})\Delta(J^{(1)})J\otimes J^{-(1)}\Lambda_{(2)}P_{n-2}(J^{(2)}J^{-(2)}\Lambda_{(3)})S(R)Q_J\\&=(Q_J^{-1}\otimes 1)((1\otimes Q_J^{-1})(S\otimes S)(J_{21}))\Delta(\Lambda_{(1)})\Delta(J^{(1)})J\end{aligned}$$

$$
\begin{aligned}
&\otimes J^{-(1)}\Lambda_{(2)}P_{n-2}(J^{(2)}J^{-(2)}\Lambda_{(3)})S(R)Q_J \quad \text{(by (8.8))}\\
&=(Q_J^{-1}\otimes 1)(1\otimes \widetilde{J}^{-(1)})\Delta(S(\widetilde{J}^{-(2)}))\Delta(\Lambda_{(1)})\Delta(J^{(1)})J\\
&\quad \otimes J^{-(1)}\Lambda_{(2)}P_{n-2}(J^{(2)}J^{-(2)}\Lambda_{(3)})S(R)Q_J \quad \text{(by (8.13))}\\
&=(Q_J^{-1}\otimes \widetilde{J}^{-(1)})\Delta(S(\widetilde{J}^{-(2)})\Lambda_{(1)})\Delta(J^{(1)})J\otimes J^{-(1)}\Lambda_{(2)}P_{n-2}(J^{(2)}J^{-(2)}\Lambda_{(3)})S(R)Q_J\\
&=(Q_J^{-1}\otimes \widetilde{J}^{-(1)})\Delta(\Lambda_{(1)})\Delta(J^{(1)})J\\
&\quad \otimes J^{-(1)}\widetilde{J}^{-(2)}_{(1)}\Lambda_{(2)}P_{n-2}(J^{(2)}J^{-(2)}\widetilde{J}^{-(2)}_{(2)}\Lambda_{(3)})S(R)Q_J \quad \text{(by (8.1))}\\
&=(Q_J^{-1}\otimes \widetilde{J}^{-(1)})\Delta(\Lambda_{(1)})(\widetilde{J}^{(1)}\otimes \widetilde{J}^{(2)}_{(1)}J^{(1)})\\
&\quad \otimes J^{-(1)}\widetilde{J}^{-(2)}_{(1)}\Lambda_{(2)}P_{n-2}(\widetilde{J}^{(2)}_{(2)}J^{(2)}J^{-(2)}\widetilde{J}^{-(2)}_{(2)}\Lambda_{(3)})S(R)Q_J \quad \text{(by (8.6))}\\
&=Q_J^{-1}\Lambda_{(1)}\widetilde{J}^{(1)}\otimes \widetilde{J}^{-(1)}\Lambda_{(2)}\widetilde{J}^{(2)}_{(1)}J^{(1)}\\
&\quad \otimes J^{-(1)}\widetilde{J}^{-(2)}_{(1)}\Lambda_{(3)}P_{n-2}(\widetilde{J}^{(2)}_{(2)}J^{(2)}J^{-(2)}\widetilde{J}^{-(2)}_{(2)}\Lambda_{(4)})S(R)Q_J.
\end{aligned}
$$

It follows that

$$
\begin{aligned}
&\Lambda_{\langle 1\rangle}\otimes \Lambda_{\langle 2\rangle}\Lambda_{\langle 3\rangle}\cdots\Lambda_{\langle n+1\rangle}\\
&=Q_J^{-1}\Lambda_{(1)}\widetilde{J}^{(1)}\otimes \widetilde{J}^{-(1)}\Lambda_{(2)}\widetilde{J}^{(2)}_{(1)}J^{(1)}J^{-(1)}\widetilde{J}^{-(2)}_{(1)}\\
&\quad \otimes \Lambda_{(3)}P_{n-2}(\widetilde{J}^{(2)}_{(2)}J^{(2)}J^{-(2)}\widetilde{J}^{-(2)}_{(2)}\Lambda_{(4)})S(R)Q_J\\
&=Q_J^{-1}\Lambda_{(1)}\widetilde{J}^{(1)}\otimes \widetilde{J}^{-(1)}\Lambda_{(2)}\widetilde{J}^{(2)}_{(1)}\widetilde{J}^{-(2)}_{(1)}\Lambda_{(3)}P_{n-2}(\widetilde{J}^{(2)}_{(2)}\widetilde{J}^{-(2)}_{(2)}\Lambda_{(4)})S(R)Q_J\\
&=Q_J^{-1}\Lambda_{(1)}\widetilde{J}^{(1)}\otimes \widetilde{J}^{-(1)}\Lambda_{(2)}P_{n-1}(\widetilde{J}^{(2)}\widetilde{J}^{-(2)}\Lambda_{(3)})S(R)Q_J\\
&=Q_J^{-1}\Lambda_{(1)}J^{(1)}\otimes J^{-(1)}\Lambda_{(2)}P_{n-1}(J^{(2)}J^{-(2)}\Lambda_{(3)})S(R)Q_J.
\end{aligned}
$$

The equality of Part (1) is now proved by induction on n. □

We have the following relationship between $P_n^J(\Lambda)$ and $P_n(\Lambda)$. This is the main result of this section.

Theorem 8.1.4 *Let H be a finite dimensional Hopf algebra over a field $\Bbbk$ with a nonzero left integral Λ and a normalized twist J. For any $n\in\mathbb{Z}$, we have*

(1) $P_n^J(\Lambda)=TP_n(\Lambda)$, where $T=J^{-(1)}\alpha(J^{-(2)})$.

(2) $P_n^J(\Lambda)=Q_J^{-1}P_n(\Lambda)S(R)Q_J$, where $R=\alpha(J^{-(1)})J^{-(2)}$.

If, moreover, H is unimodular, then $P_n^J(\Lambda)=P_n(\Lambda)$.

Proof (1) It is direct to check that the identity in question holds for the case $n=0,\pm1$. For the case $n\geqslant 2$, we have

$$
\begin{aligned}
P_n^J(\Lambda)&=\Lambda_{\langle 1\rangle}\Lambda_{\langle 2\rangle}\Lambda_{\langle 3\rangle}\cdots\Lambda_{\langle n\rangle}\\
&=Q_J^{-1}\Lambda_{(1)}J^{(1)}J^{-(1)}\Lambda_{(2)}P_{n-2}(J^{(2)}J^{-(2)}\Lambda_{(3)})S(R)Q_J \quad \text{(by Lemma 8.1.3 (1))}
\end{aligned}
$$

$$
\begin{aligned}
&= Q_J^{-1}\Lambda_{(1)}\Lambda_{(2)}P_{n-2}(\Lambda_{(3)})S(R)Q_J \\
&= Q_J^{-1}P_n(\Lambda)S(R)Q_J \\
&= P_n(\Lambda)J^{(1)}(\alpha\circ S^{-1})(J^{(2)}Q_J^{-1}) \quad \text{(by Proposition 8.1.1)} \\
&= P_n(\Lambda)J_{(1)}^{-(1)}(\alpha\circ S^{-1})(J_{(2)}^{-(1)})\alpha(J^{-(2)}) \quad \text{(by (8.10))} \\
&= TP_n(\Lambda) \quad \text{(by Proposition 8.1.1)}.
\end{aligned}
$$

For the case $n \leqslant -2$, we have

$$
\begin{aligned}
P_n^J(\Lambda) &= S^J(\Lambda_{\langle -n\rangle}\cdots\Lambda_{\langle 1\rangle}) \\
&= Q_J^{-1}S(\Lambda_{\langle -n\rangle}\cdots\Lambda_{\langle 1\rangle})Q_J \\
&= Q_J^{-1}S(S^{-1}(Q_J^{-1})\Lambda_{(3)}J^{(2)}J^{-(2)}\Lambda_{(2)}(S^{-1}\circ P_{n+2}) \\
&\quad \times (J^{(1)}J^{-(1)}\Lambda_{(1)})S^{-1}(T)S^{-1}(Q_J))Q_J \\
&= Q_J^{-1}S(S^{-1}(Q_J^{-1})\Lambda_{(3)}\Lambda_{(2)}(S^{-1}\circ P_{n+2})(\Lambda_{(1)})S^{-1}(T)S^{-1}(Q_J))Q_J \\
&= S(\Lambda_{(3)}\Lambda_{(2)}(S^{-1}\circ P_{n+2})(\Lambda_{(1)})S^{-1}(T)) \\
&= TP_n(\Lambda),
\end{aligned}
$$

where the third equality follows from Lemma 8.1.3 (2).

(2) On the one hand,

$$
\begin{aligned}
P_n^J(\Lambda) &= TP_n(\Lambda) \\
&= P_n(\Lambda)J_{(1)}^{-(1)}(\alpha\circ S^{-1})(J_{(2)}^{-(1)})\alpha(J^{-(2)}) \quad \text{(by Proposition 8.1.1)} \\
&= P_n(\Lambda)J^{(1)}(\alpha\circ S^{-1})(J^{(2)}Q_J^{-1}) \quad \text{(by (8.10))}.
\end{aligned}
$$

On the other hand,

$$
\begin{aligned}
&Q_J^{-1}P_n(\Lambda)S(R)Q_J \\
&= P_n(\Lambda)(Q_J^{-1})_{(1)}(\alpha\circ S^{-1})((Q_J^{-1})_{(2)})S(R)Q_J \quad \text{(by Proposition 8.1.1)} \\
&= P_n(\Lambda)J^{(1)}Q_J^{-1}S(\widetilde{J}^{(2)})(\alpha\circ S^{-1})(J^{(2)}Q_J^{-1}S(\widetilde{J}^{(1)}))S(R)Q_J \quad \text{(by (8.8))} \\
&= P_n(\Lambda)J^{(1)}(\alpha\circ S^{-1})(J^{(2)}Q_J^{-1}).
\end{aligned}
$$

We conclude that

$$
P_n^J(\Lambda) = Q_J^{-1}P_n(\Lambda)S(R)Q_J, \quad \text{for any } n\in\mathbb{Z}.
$$

If H is unimodular, then $\alpha=\varepsilon$ and hence $T=1$, this implies that

$$
P_n^J(\Lambda) = P_n(\Lambda), \quad \text{for any } n\in\mathbb{Z}. \qquad \square
$$

Remark 8.1.5 *Theorem* 8.1.4 *shows that the sequence* $\{P_n(\Lambda)\}_{n\in\mathbb{Z}}$ *of a unimodular Hopf algebra is invariant under twisting.*

8.2 Polynomial invariants

In this section, we will use Theorem 8.1.4 to give certain polynomial invariants of representation categories of Hopf algebras. As applications, we show that the representation categories of 12-dimensional pointed nonsemisimple Hopf algebras classified in [59] are mutually inequivalent as monoidal categories. We also show that the 8-dimensional semisimple Hopf algebras, including Kac algebra K_8, the dihedral group algebra $\Bbbk D_4$ and the quaternion group algebra $\Bbbk Q_8$ are mutually inequivalent as monoidal categories.

We begin with the following preparations. Let H and H' be finite dimensional Hopf algebras over a field $\Bbbk$ with nonzero left integrals Λ and Λ' respectively. If the functor $\mathcal{F}: H\text{-mod} \to H'\text{-mod}$ is an equivalence as monoidal categories, it follows from [63, Theorem 2.2] that there exists a normalized twist J of H such that H' is isomorphic to H^J as bialgebras. Let $\sigma : H' \to H^J$ be such an isomorphism. Then σ is automatically a Hopf algebra isomorphism. Therefore, for any $n \in \mathbb{Z}$, we have

$$\sigma \circ P'_n = P^J_n \circ \sigma,$$

where P'_n and P^J_n are the n-th Sweedler power maps of H' and H^J respectively. Suppose $\sigma(\Lambda') = \mu\Lambda$ for a nonzero scalar $\mu \in \Bbbk$. Then

$$\sigma(P'_n(\Lambda')) = P^J_n(\sigma(\Lambda')) = \mu P^J_n(\Lambda) = \mu T P_n(\Lambda), \tag{8.17}$$

where $T = J^{-(1)}\alpha(J^{-(2)})$. The isomorphism σ induces a $\Bbbk$-linear equivalence $(-)^\sigma$: H-mod $\to$ H'-mod as follows: for any finite dimensional H-module V, $V^\sigma = V$ as $\Bbbk$-linear space with the H'-module structure given by $a'v = \sigma(a')v$ for $a' \in H'$, $v \in V$, and $f^\sigma = f$ for any morphism f in H-mod. Thus,

$$\chi_{V^\sigma}(a') = \chi_V(\sigma(a')), \quad \text{for } a' \in H'.$$

Moreover, the equivalence $\mathcal{F}$ is naturally isomorphic to the $\Bbbk$-linear equivalence $(-)^\sigma$ (see [42, Theorem 1.1]). Therefore,

$$\chi_{\mathcal{F}(V)}(a') = \chi_{V^\sigma}(a'), \quad \text{for } a' \in H'.$$

By taking a' to be $P'_n(\Lambda')$, we have

$$\chi_{\mathcal{F}(V)}(P'_n(\Lambda')) = \chi_{V^\sigma}(P'_n(\Lambda')) = \chi_V(\sigma(P'_n(\Lambda'))) = \mu\chi_V(TP_n(\Lambda)), \tag{8.18}$$

where the scalar μ is independent on the choice of V.

The first result of this section states that any homogeneous polynomial vanishing on some $P_n(\Lambda)$ is a monoidal invariant of the representation category H-mod. We shall see that this invariant can be used to distinguish the representation categories of some Hopf algebras.

Theorem 8.2.1 *Let H and H' be finite dimensional Hopf algebras over a field $\Bbbk$ with nonzero left integrals Λ and Λ' respectively. Let H-mod and H'-mod be equivalent as $\Bbbk$-linear monoidal categories. For any homogeneous polynomial $\psi(X_1, \cdots, X_s) \in \Bbbk[X_1, \cdots, X_s]$ and $n_1, \cdots, n_s \in \mathbb{Z}$, we have*

$$\psi(P_{n_1}(\Lambda), \cdots, P_{n_s}(\Lambda)) = 0 \text{ if and only if } \psi(P'_{n_1}(\Lambda'), \cdots, P'_{n_s}(\Lambda')) = 0.$$

Proof Since H-mod and H'-mod are equivalent as monoidal categories, there exists a normalized twist J of H such that H' is isomorphic to H^J as Hopf algebras. Let $\sigma : H' \to H^J$ be such an isomorphism. It follows from (8.17) that

$$\sigma(P'_n(\Lambda')) = \mu T P_n(\Lambda),$$

where μ is a nonzero scalar and $T = J^{-(1)}\alpha(J^{-(2)})$ which is an invertible element of H. Let $\psi(X_1, \cdots, X_s)$ be a homogeneous polynomial of degree m. Then

$$\begin{aligned}\sigma(\psi(P'_{n_1}(\Lambda'), \cdots, P'_{n_s}(\Lambda'))) &= \psi(\sigma(P'_{n_1}(\Lambda')), \cdots, \sigma(P'_{n_s}(\Lambda')))\\ &= \psi(\mu T P_{n_1}(\Lambda), \cdots, \mu T P_{n_s}(\Lambda))\\ &= \psi(P_{n_1}(\Lambda), \cdots, P_{n_s}(\Lambda))\mu^m T_m \cdots T_2 T_1 \quad \text{(by (8.15))},\end{aligned}$$

where these elements $T_1, \cdots, T_m \in H$ are as defined recursively in (8.14). Note that these elements $T_1, \cdots, T_m$ are invertible in H. Therefore, $\psi(P_{n_1}(\Lambda), \cdots, P_{n_s}(\Lambda)) = 0$ if and only if $\psi(P'_{n_1}(\Lambda'), \cdots, P'_{n_s}(\Lambda')) = 0$. The proof is completed. □

Note that the (not necessarily homogeneous) polynomials vanishing on some $P_n(\Lambda)$ are in general not anymore invariants of the representation category H-mod. But for a unimodular Hopf algebra H, as shown below that these polynomials are still something meaningful to distinguish the representation category H-mod.

Proposition 8.2.2 *Let H and H' be finite dimensional unimodular Hopf algebras over a field $\Bbbk$. Let H-mod and H'-mod be equivalent as $\Bbbk$-linear monoidal categories. For any polynomial $\psi(X_1, \cdots, X_s) \in \Bbbk[X_1, \cdots, X_s]$, if $\psi(P_{n_1}(\Lambda), \cdots, P_{n_s}(\Lambda)) = 0$ for a nonzero left integral $\Lambda \in H$ and some $n_1, \cdots, n_s \in \mathbb{Z}$, then there exists a nonzero left integral $\Lambda' \in H'$ such that $\psi(P'_{n_1}(\Lambda'), \cdots, P'_{n_s}(\Lambda')) = 0$.*

Proof Since H-mod and H'-mod are equivalent as monoidal categories, there exists a normalized twist J of H such that H' is isomorphic via a map σ to H^J as Hopf algebras. We may choose a left integral $\Lambda' \in H'$ such that $\sigma(\Lambda') = \Lambda$. In this case, the scalar μ appeared in (8.17) is 1. Since H is unimodular, the element T

appeared in (8.17) is also 1. Now (8.17) has the form $\sigma(P'_n(\Lambda')) = P_n(\Lambda)$ for any $n \in \mathbb{Z}$. Thus, $\psi(P_{n_1}(\Lambda), \cdots, P_{n_s}(\Lambda)) = 0$ implies that $\psi(P'_{n_1}(\Lambda'), \cdots, P'_{n_s}(\Lambda')) = 0$. The proof is completed. □

For a semisimple Hopf algebra H with an idempotent integral Λ, the following result shows that the polynomial with several variables vanishing on some $P_n(\Lambda)$ is a monoidal invariant of the representation category H-mod.

Proposition 8.2.3 *Let H and H' be semisimple Hopf algebras over a field $\Bbbk$ with idempotent integrals Λ and Λ' respectively. Let H-mod and H'-mod be equivalent as $\Bbbk$-linear monoidal categories. We have $\psi(P_{n_1}(\Lambda), \cdots, P_{n_s}(\Lambda)) = 0$ if and only if $\psi(P'_{n_1}(\Lambda'), \cdots, P'_{n_s}(\Lambda')) = 0$, where $\psi(X_1, \cdots, X_s) \in \Bbbk[X_1, \cdots, X_s]$ and $n_1, \cdots, n_s \in \mathbb{Z}$.*

Proof Since H-mod and H'-mod are equivalent as monoidal categories, there exists a normalized twist J of H such that H' is isomorphic via a map σ to H^J as Hopf algebras. Since Λ and Λ' are both idempotent and A is unimodular, the scalar μ and the element T appeared in (8.17) are both equal to 1. Now (8.17) becomes $\sigma(P'_n(\Lambda')) = P_n(\Lambda)$ for any $n \in \mathbb{Z}$. Thus, $\psi(P_{n_1}(\Lambda), \cdots, P_{n_s}(\Lambda)) = 0$ if and only if $\psi(P'_{n_1}(\Lambda'), \cdots, P'_{n_s}(\Lambda')) = 0$. The proof is completed. □

Although these $P_n(\Lambda)$ for $n \in \mathbb{Z}$ are invariant under twisting for a finite dimensional unimodular Hopf algebra H, it is clear that $P_n(\Lambda)$ are dependent on the choice of Λ, and the values that characters of finite dimensional H-modules taking on them are not monoidal invariants in general. Note that if H is unimodular then the element T appeared in (8.18) is the identity 1. Now (8.18) has the form

$$\chi_{\mathcal{F}(V)}(P'_n(\Lambda')) = \mu\chi_V(P_n(\Lambda)), \quad \text{for } n \in \mathbb{Z},$$

where μ is a nonzero scalar which is independent on the choice of V. In view of this, if $\chi_W(P_m(\Lambda)) \neq 0$ for some finite dimensional H-module W and some $m \in \mathbb{Z}$, then for any finite dimensional H-module V and $n \in \mathbb{Z}$, we have

$$\chi_{\mathcal{F}(V)}(P'_n(\Lambda'))/\chi_{\mathcal{F}(W)}(P'_m(\Lambda')) = \chi_V(P_n(\Lambda))/\chi_W(P_m(\Lambda)).$$

That is, the ratio $\chi_V(P_n(\Lambda))/\chi_W(P_m(\Lambda))$ is a monoidal invariant of the representation category H-mod. We summarize it as follows:

Proposition 8.2.4 *Let H be a finite dimensional unimodular Hopf algebra over a field $\Bbbk$ with a nonzero left integral Λ. If $\chi_W(P_m(\Lambda)) \neq 0$ for some finite dimensional H-module W and $m \in \mathbb{Z}$, then for any finite dimensional H-module V and $n \in \mathbb{Z}$, the ratio $\chi_V(P_n(\Lambda))/\chi_W(P_m(\Lambda))$ is a monoidal invariant of H-mod.*

For a semisimple Hopf algebra H over a field $\Bbbk$ with an idempotent integral Λ, the value $\chi_V(P_n(\Lambda))$ is called the n-th Frobenius-Schur indicator of a finite dimensional H-module V (see [43, 50]). It follows from [63, Proposition 3.2] that $\chi_V(P_n(\Lambda))$ is a monoidal invariant of H-mod. Proposition 8.2.4 can be regarded as a slightly generalization of the n-th Frobenius-Schur indicator from a semisimple Hopf algebra to a unimodular Hopf algebra. Indeed, if H is semisimple and W is chosen to be the trivial H-module $\Bbbk$, then $\chi_{\Bbbk}(P_m(\Lambda)) = \varepsilon(\Lambda) \neq 0$ and the ratio

$$\chi_V(P_n(\Lambda))/\chi_{\Bbbk}(P_m(\Lambda)) = \chi_V(P_n(\Lambda/\varepsilon(\Lambda))),$$

which is the n-th Frobenius-Schur indicator of V for any $n \in \mathbb{Z}$. Note that the base field $\Bbbk$ is arbitrary. It suggests that the invariant of the n-th Frobenius-Schur indicator of V is valid for a semisimple Hopf algebra over an arbitrary field $\Bbbk$.

For a finite dimensional non-semisimple Hopf algebra H, the following result suggests that it is interesting to seek for an H-module W such that $\chi_W(P_m(\Lambda)) \neq 0$ for some $m \in \mathbb{Z}$.

Proposition 8.2.5 *Let H be a finite dimensional Hopf algebra over a field $\Bbbk$ with a nonzero left integral $\Lambda \in H$ and the distinguished group-like element $\alpha \in H^*$. Assume in addition that one of the following conditions is satisfied:*

(1) *H is non-semisimple with the Chevalley property.*

(2) *$\alpha(a) \neq 1$ for some group-like element $a \in H$.*

(3) *$S^2(\Lambda) \neq \Lambda$.*

Then $\chi_V(P_n(\Lambda)) = 0$ for any finite dimensional H-module V and $n \in \mathbb{Z}$.

Proof (1) Since H is a non-semisimple Hopf algebra, any nonzero left integral Λ is a nilpotent element of H. It follows that Λ belongs to the Jacobson radical $\mathrm{rad}(H)$ of H. Note that the Hopf algebra H has the Chevalley property. Namely, the Jacobson radical $\mathrm{rad}(H)$ is a Hopf ideal of H. Therefore, $P_n(\Lambda) \in \mathrm{rad}(H)$ and hence $\chi_V(P_n(\Lambda)) = 0$ for any finite dimensional H-module V and $n \in \mathbb{Z}$.

(2) It follows from Proposition 8.1.1 that

$$a^{-1}P_n(\Lambda)a = \alpha(a)P_n(\Lambda),$$

for any group-like element a of H. Therefore, if $\alpha(a) \neq 1$ for some group-like element a, then $\chi_V(P_n(\Lambda)) = 0$ for any finite dimensional H-module V and $n \in \mathbb{Z}$.

(3) Note that

$$S^2(\Lambda) = \alpha(g_0^{-1})\Lambda,$$

where g_0 is the distinguished group-like element of H (see [70, Proposition 3 (d)]). Therefore, if $S^2(\Lambda) \neq \Lambda$, then $\alpha(g_0^{-1}) \neq 1$ and hence $\chi_V(P_n(\Lambda)) = 0$ by Part (2). □

At the end of this section, we pay a little attention to the Hopf order of a nonzero left integral of H. Recall from [45] that the *Hopf order* of a nonzero left integral $\Lambda \in H$ is the least positive integer n such that $P_n(\Lambda) = P_0(\Lambda)$. If such n does not exist, the Hopf order of Λ is infinity. If H and H' are finite dimensional Hopf algebras with nonzero left integrals Λ and Λ' respectively such that H-mod and H'-mod are equivalent as $\Bbbk$-linear monoidal categories, then $\sigma(P_n'(\Lambda')) = \mu T P_n(\Lambda)$

by (8.17). It follows that $P'_n(\Lambda') = P'_0(\Lambda')$ if and only if $P_n(\Lambda) = P_0(\Lambda)$. Therefore, the Hopf order of Λ is equal to that of Λ'. That is, the Hopf order of a nonzero left integral $\Lambda \in H$ is a monoidal invariant of H-mod. It can be seen from the next section that the Hopf orders of $\Lambda \in \Bbbk D_4$ and $\Lambda \in \Bbbk Q_8$ are both equal to 4, while the Hopf order of $\Lambda \in K_8$ is equal to 8.

8.3 Examples

In this section, we first use Theorem 8.2.1 to distinguish the representation categories of 12-dimensional pointed non-semisimple Hopf algebras.

Example 8.3.1 The following table gives a list of pairwise non-isomorphic pointed non-semisimple Hopf algebras of dimension 12 over an algebraically closed field $\Bbbk$ of characteristic zero. Every Hopf algebra in the list is presented by two generators g and x subject to the following relations:

$$\mathcal{A}_0: \quad g^6 = 1,\ x^2 = 0,\ gx = -xg,\ \Delta(g) = g \otimes g,\ \Delta(x) = x \otimes 1 + g \otimes x,$$

$$\mathcal{A}_1: \quad g^6 = 1,\ x^2 = 1 - g^2,\ gx = -xg,\ \Delta(g) = g \otimes g,\ \Delta(x) = x \otimes 1 + g \otimes x,$$

$$\mathcal{B}_0: \quad g^6 = 1,\ x^2 = 0,\ gx = -xg,\ \Delta(g) = g \otimes g,\ \Delta(x) = x \otimes 1 + g^3 \otimes x,$$

$$\mathcal{B}_1: \quad g^6 = 1,\ x^2 = 0,\ gx = \omega xg,\ \Delta(g) = g \otimes g,\ \Delta(x) = x \otimes 1 + g^3 \otimes x,$$

where $\omega \in \Bbbk$ is a fixed primitive 6-th root of unity. It follows from [59] that these four Hopf algebras are up to isomorphism all pointed non-semisimple Hopf algebras of dimension 12. These Hopf algebras are not unimodular and have nonzero left integrals of the same form

$$\Lambda = (1 + g + g^2 + g^3 + g^4 + g^5)x.$$

Since $S^2(\Lambda) = -\Lambda$ holds for all these Hopf algebras, it follows from [72, Proposition 3.13] that the (-1)-th indicators ν_{-1} of these Hopf algebras are all -1. For the trace of antipode S of each Hopf algebra, we have $\mathrm{tr}(S) = 2$. Hence the monoidal invariants ν_{-1} and $\nu_2 = \mathrm{tr}(S)$ which are easy to handle can not be used to distinguish the representation categories of these Hopf algebras. However, we may find some homogeneous polynomials of degree 1 vanishing on some $P_n(\Lambda)$ as follows:

$\mathcal{A}_0$	$P_2(\Lambda) + P_{-2}(\Lambda) = 0$	$P_3(\Lambda) - 3P_2(\Lambda) - P_{-3}(\Lambda) = 0$
$\mathcal{A}_1$	$P_2(\Lambda) + P_{-2}(\Lambda) = 0$	$P_3(\Lambda) - 3P_2(\Lambda) - P_{-3}(\Lambda) = 0$
$\mathcal{B}_0$	$P_2(\Lambda) + P_{-2}(\Lambda) = 0$	$P_3(\Lambda) - 3P_2(\Lambda) - P_{-3}(\Lambda) \neq 0$
$\mathcal{B}_1$	$P_2(\Lambda) + P_{-2}(\Lambda) \neq 0$	

According to Theorem 8.2.1, we may see from the second column of the above table that $\mathcal{B}_1$-mod is not monoidally equivalent to $\mathcal{A}_0$-mod, $\mathcal{A}_1$-mod and $\mathcal{B}_0$-mod.

Similarly, it can be seen from the third column that $\mathcal{B}_0$-mod is not monoidally equivalent to $\mathcal{A}_0$-mod and $\mathcal{A}_1$-mod. However, this approach can not be used to distinguish the representation categories $\mathcal{A}_0$-mod and $\mathcal{A}_1$-mod, since $P_n(\Lambda) \in \mathcal{A}_0$ and $P_n(\Lambda) \in \mathcal{A}_1$ have the same expression for any $n \in \mathbb{Z}$ with respect to the $\Bbbk$-basis $\{g^i x^j \mid 0 \leqslant i \leqslant 5, 0 \leqslant j \leqslant 1\}$ of $\mathcal{A}_0$ and of $\mathcal{A}_1$. Fortunately, the representation categories of $\mathcal{A}_0$ and $\mathcal{A}_1$ have already been investigated in [86] and [87] respectively. The number of finite dimensional indecomposable representations of $\mathcal{A}_0$ up to isomorphism is 12 (see [86, Theorem 2.5]), while the number for that of $\mathcal{A}_1$ is 6 (see [87, Theorem 2.9]). In summary, the representation categories of these Hopf algebras $\mathcal{A}_0$, $\mathcal{A}_1$, $\mathcal{B}_0$ and $\mathcal{B}_1$ are mutually inequivalent as monoidal categories.

In the sequel, we will use Proposition 8.2.3 to distinguish the representation categories $\Bbbk Q_8$-mod, $\Bbbk D_4$-mod and K_8-mod over a field $\Bbbk$ of char$(\Bbbk) \neq 2$. Note that these representation categories have the same Grothendieck ring but they are inequivalent as monoidal categories as they have different higher Frobenius-Schur indicators (see [63, Theorem 6.1]).

Example 8.3.2 The quaternion group Q_8 is a group with eight elements, which can be described in the following way: It is the group formed by eight elements $\mathbb{1}, -\mathbb{1}, i, -i, j, -j, k, -k$ where $\mathbb{1}$ is the identity element, $(-\mathbb{1})^2 = \mathbb{1}$ and all the other elements are squareroots of $-\mathbb{1}$, such that $(-\mathbb{1})i = -i, (-\mathbb{1})j = -j, (-\mathbb{1})k = -k$ and further, $ij = k, ji = -k, jk = i, kj = -i, ki = j, ik = -j$ (the remaining relations can be deduced from these). The group algebra $\Bbbk Q_8$ has the idempotent integral

$$\Lambda = \frac{1}{8}(\mathbb{1} + (-\mathbb{1}) + i + (-i) + j + (-j) + k + (-k)).$$

By a straightforward computation, we have

$$P_1(\Lambda) = \Lambda, \quad P_2(\Lambda) = \frac{1}{4}\mathbb{1} + \frac{3}{4}(-\mathbb{1}), \quad P_3(\Lambda) = P_1(\Lambda), \quad P_4(\Lambda) = P_0(\Lambda).$$

Note that

$$P_4(\Lambda) - P_0(\Lambda) = 0 \quad \text{and} \quad 2(P_2(\Lambda))^2 - P_2(\Lambda) - P_0(\Lambda) = 0.$$

For the dihedral group

$$D_4 = \{1, a, a^2, a^3, b, ba, ba^2, ba^3\}$$

with $a^4 = 1$, $b^2 = 1$ and $aba = b$, the group algebra $\Bbbk D_4$ has the idempotent integral

$$\Lambda = \frac{1}{8}(1 + a + a^2 + a^3 + b + ba + ba^2 + ba^3).$$

We have

$$P_1(\Lambda) = \Lambda, \quad P_2(\Lambda) = \frac{3}{4} + \frac{1}{4}a^2, \quad P_3(\Lambda) = P_1(\Lambda), \quad P_4(\Lambda) = P_0(\Lambda).$$

Note that $P_4(\Lambda) - P_0(\Lambda) = 0$ while

$$2(P_2(\Lambda))^2 - P_2(\Lambda) - P_0(\Lambda) = \frac{1}{2}a^2 - \frac{1}{2} \neq 0$$

The 8-dimensional Kac algebra K_8 is a semisimple Hopf algebra over $\Bbbk$ generated by x, y, z as a $\Bbbk$-algebra with the following relations (see [53]):

$$x^2 = y^2 = 1, \quad z^2 = \frac{1}{2}(1 + x + y - xy), \quad xy = yx, \quad xz = zy, \quad yz = zx.$$

The coalgebra structure Δ, ε and the antipode S of K_8 are given by

$$\Delta(x) = x \otimes x, \quad \Delta(y) = y \otimes y, \quad \varepsilon(x) = \varepsilon(y) = 1,$$

$$\Delta(z) = \frac{1}{2}(1 \otimes 1 + 1 \otimes x + y \otimes 1 - y \otimes x)(z \otimes z), \quad \varepsilon(z) = 1,$$

$$S(x) = x, \quad S(y) = y, \quad S(z) = z.$$

The idempotent integral of K_8 is

$$\Lambda = \frac{1}{8}(1 + x + y + xy) + \frac{1}{8}(1 + x + y + xy)z.$$

A straightforward computation shows that

$$P_1(\Lambda) = \Lambda, \quad P_2(\Lambda) = \frac{3}{4} + \frac{1}{4}xy, \quad P_3(\Lambda) = P_1(\Lambda), \quad P_4(\Lambda) = \frac{1}{2} + \frac{1}{2}xy,$$

$$P_5(\Lambda) = P_1(\Lambda), \quad P_6(\Lambda) = P_2(\Lambda), \quad P_7(\Lambda) = P_1(\Lambda), \quad P_8(\Lambda) = P_0(\Lambda).$$

Note that

$$P_4(\Lambda) - P_0(\Lambda) = \frac{1}{2}xy - \frac{1}{2} \neq 0.$$

We may see from the following table that for any two of these Hopf algebras, there exists a polynomial ψ in three variables such that

$$\psi(P_0(\Lambda), P_2(\Lambda), P_4(\Lambda)) = 0$$

holds for one Hopf algebra but not for another Hopf algebra. Thus, the three representation categories are mutually inequivalent as monoidal categories by Proposition 8.2.3.

K_8	$P_4(\Lambda) - P_0(\Lambda) \neq 0$		$P_4(\Lambda) - P_0(\Lambda) \neq 0$
$\Bbbk D_4$	$P_4(\Lambda) - P_0(\Lambda) = 0$	$2(P_2(\Lambda))^2 - P_2(\Lambda) - P_0(\Lambda) \neq 0$	
$\Bbbk Q_8$		$2(P_2(\Lambda))^2 - P_2(\Lambda) - P_0(\Lambda) = 0$	$P_4(\Lambda) - P_0(\Lambda) = 0$

We finally give an example of a non-semisimple Hopf algebra in which $\chi_V(P_n(\Lambda)) \neq 0$ for some integer n and some representation V.

Example 8.3.3 We assume that the ground field $\Bbbk$ is the field of complex numbers. Let q be a primitive 4-th root of unity. We consider a restricted quantum group $\overline{U}_q(sl_2)$ of dimension 16. By definition, $\overline{U}_q(sl_2)$ is an associative $\Bbbk$-algebra with the unity 1 generated by E, F, K, K^{-1} subject to the relations

$$KK^{-1} = K^{-1}K = 1, \quad KEK^{-1} = q^2E, \quad KFK^{-1} = q^{-2}F,$$

$$EF - FE = \frac{K - K^{-1}}{q - q^{-1}}, \quad K^4 = 1, \quad E^2 = F^2 = 0.$$

It is a Hopf algebra with comultiplication Δ, counit ε, and antipode S given by

$$\Delta(K) = K \otimes K, \quad \Delta(K^{-1}) = K^{-1} \otimes K^{-1},$$

$$\Delta(E) = 1 \otimes E + E \otimes K, \quad \Delta(F) = K^{-1} \otimes F + F \otimes 1,$$

$$\varepsilon(K) = \varepsilon(K^{-1}) = 1, \quad \varepsilon(E) = \varepsilon(F) = 0,$$

$$S(K) = K^{-1}, \quad S(K^{-1}) = K, \quad S(E) = -EK^{-1}, \quad S(F) = -KF.$$

Obviously, $\overline{U}_q(sl_2)$ is a restricted quantum group of dimension 16 with a $\Bbbk$-basis $\{F^jE^lK^i \mid 0 \leqslant j, l \leqslant 1, 0 \leqslant i \leqslant 3\}$ and $\Lambda = FE(\sum_{i=0}^{3} K^i)$ is a nonzero left and right integral of $\overline{U}_q(sl_2)$ (see [31, Eq.(3.2)]). We have

$$\begin{aligned}\Delta(\Lambda) &= \Lambda_{(1)} \otimes \Lambda_{(2)}\\ &= (K^{-1} \otimes F + F \otimes 1)(1 \otimes E + E \otimes K)\sum_{i=0}^{3}(K^i \otimes K^i)\\ &= \sum_{i=0}^{3}(K^{i-1} \otimes FEK^i + K^{-1}EK^i \otimes FK^{i+1} + FK^i \otimes EK^i + FEK^i \otimes K^{i+1})\end{aligned}$$

and

$$\begin{aligned}P_2(\Lambda) &= \Lambda_{(1)}\Lambda_{(2)}\\ &= \sum_{i=0}^{3}(K^{i-1}FEK^i + K^{-1}EK^iFK^{i+1} + FK^iEK^i + FEK^iK^{i+1})\end{aligned}$$

$$= \sum_{i=0}^{3}(FEK^{2i-1} + q^{-2i}EFK^{2i} + q^{2i}FEK^{2i} + FEK^{2i+1}).$$

Recall from [31, Section 3.2.1] that the Hopf algebra $\overline{U}_q(sl_2)$ has a simple module V of dimension 2 with a basis $\{v_0, v_1\}$, where v_0 is the highest weight vector and the actions of the generators K, E and F on V with respect to this basis are given by

$$\begin{cases} Kv_0 = qv_0, \\ Kv_1 = q^{-1}v_1, \end{cases} \qquad \begin{cases} Ev_0 = 0, \\ Ev_1 = v_0, \end{cases} \qquad \begin{cases} Fv_0 = v_1, \\ Fv_1 = 0. \end{cases}$$

It follows that

$$\begin{cases} P_2(\Lambda)v_0 = 4v_0, \\ P_2(\Lambda)v_1 = 4v_1. \end{cases}$$

Therefore, $\chi_V(P_2(\Lambda)) = 8$.

8.4 Integrals of the dual of twisted Hopf algebras

In this section, we investigate the relationship between a right integral $\lambda \in H^*$ and a right integral $\lambda^J \in (H^J)^*$. Based on this investigation, we provide a unifying proof of the well-known result which says that the indicator $\nu_n(H)$ of a finite dimensional Hopf algebra H is a monoidal invariant of H-mod for any $n \in \mathbb{Z}$. We also use the expression of λ^J to give a different proof of the known result that the Killing form of a finite dimensional Hopf algebra H is invariant under twisting.

Note that the expression of a right integral $\lambda^J \in (H^J)^*$ has been described in [1, Theorem 3.4] when H is a finite dimensional unimodular Hopf algebra. For a general Hopf algebra H, a right integral $\lambda^J \in (H^J)^*$ can be described as follows:

Theorem 8.4.1 *Let H be a finite dimensional Hopf algebra over a field $\Bbbk$ with a normalized twist J. Let $R = \alpha(J^{-(1)})J^{-(2)}$. If λ is a nonzero right integral of H^*, then*

$$\lambda^J := \lambda \leftharpoonup S^2(R^{-1})S(Q_J^{-1})Q_J$$

is a nonzero right integral of $(H^J)^*$.

Proof We need to show that

$$\lambda^J(b_{\langle 1\rangle})b_{\langle 2\rangle} = \lambda^J(b), \quad \text{for all } b \in H.$$

We set $\widetilde{J}^{(1)} \otimes \widetilde{J}^{(2)} = J^{(1)} \otimes J^{(2)}$. Then

$$\lambda^J(b_{\langle 1\rangle})b_{\langle 2\rangle}$$
$$= \lambda^J(J^{-(1)}b_{(1)}J^{(1)})J^{-(2)}b_{(2)}J^{(2)}$$

$$= \lambda(\alpha(\widetilde{J}^{(1)})S^2(\widetilde{J}^{(2)})S(Q_J^{-1})Q_JJ^{-(1)}b_{(1)}J^{(1)})J^{-(2)}b_{(2)}J^{(2)}$$
$$= \lambda(\alpha(J^{(1)}_{(1)})S^2(J^{(1)}_{(2)})\alpha(\widetilde{J}^{(1)})S^2(\widetilde{J}^{(2)})S(Q_J^{-1})Q_JJ^{-(1)}b_{(1)})J^{-(2)}b_{(2)}J^{(2)} \quad \text{(by (8.3))}$$
$$= \lambda(\alpha(J^{(1)}_{(1)}\widetilde{J}^{(1)})S^2(J^{(1)}_{(2)}\widetilde{J}^{(2)})S(Q_J^{-1})Q_JJ^{-(1)}b_{(1)})J^{-(2)}b_{(2)}J^{(2)}$$
$$= \lambda(\alpha(J^{(1)})S^2(J^{(2)}_{(1)}\widetilde{J}^{(1)})S(Q_J^{-1})Q_JJ^{-(1)}b_{(1)})J^{-(2)}b_{(2)}J^{(2)}_{(2)}\widetilde{J}^{(2)} \quad \text{(by (8.6))}.$$

We denote $t := S(Q_J^{-1})Q_J$. It follows from (8.7) and (8.8) that

$$\Delta(t) = t_{(1)} \otimes t_{(2)} = (S^2 \otimes S^2)(J)(S(Q_J^{-1})Q_J \otimes S(Q_J^{-1})Q_J)J^{-1}.$$

Applying $id \otimes S^{-1}$ to both sides of this equality, we obtain that

$$t_{(1)} \otimes S^{-1}(t_{(2)}) = S^2(\widetilde{J}^{(1)})S(Q_J^{-1})Q_JJ^{-(1)} \otimes S^{-1}(J^{-(2)})S^{-1}(Q_J)Q_J^{-1}S(\widetilde{J}^{(2)}). \tag{8.19}$$

Now we have

$$\lambda^J(b_{\langle 1\rangle})b_{\langle 2\rangle}$$
$$= \alpha(J^{(1)})J^{-(2)}\lambda(S^2(J^{(2)}_{(1)}\widetilde{J}^{(1)})tJ^{-(1)}b_{(1)})b_{(2)}J^{(2)}_{(2)}\widetilde{J}^{(2)}$$
$$= \alpha(J^{(1)})J^{-(2)}\lambda(S^2(J^{(2)}_{(1)}\widetilde{J}^{(1)}_{(1)})t_{(1)}J^{-(1)}_{(1)}b)S^{-1}$$
$$\times (S^2(J^{(2)}_{(2)}\widetilde{J}^{(1)}_{(2)})t_{(2)}J^{-(1)}_{(2)})J^{(2)}_{(3)}\widetilde{J}^{(2)} \quad \text{(by (8.4))}$$
$$= \alpha(J^{(1)})J^{-(2)}\lambda(S^2(J^{(2)}_{(1)}\widetilde{J}^{(1)}_{(1)})t_{(1)}J^{-(1)}_{(1)}b)S^{-1}(J^{-(1)}_{(2)})S^{-1}(t_{(2)})S(\widetilde{J}^{(1)}_{(2)})S(J^{(2)}_{(2)})J^{(2)}_{(3)}\widetilde{J}^{(2)}$$
$$= \alpha(J^{(1)})J^{-(2)}\lambda(S^2(J^{(2)}\widetilde{J}^{(1)}_{(1)})t_{(1)}J^{-(1)}_{(1)}b)S^{-1}(J^{-(1)}_{(2)})S^{-1}(t_{(2)})S(\widetilde{J}^{(1)}_{(2)})\widetilde{J}^{(2)}$$
$$= \lambda(S^2(R^{-1})S^2(\widetilde{J}^{(1)}_{(1)})t_{(1)}J^{-(1)}_{(1)}b)J^{-(2)}S^{-1}(J^{-(1)}_{(2)})S^{-1}(t_{(2)})S(\widetilde{J}^{(1)}_{(2)})\widetilde{J}^{(2)}$$
$$= \lambda(S^2(R^{-1})S^2(\widetilde{J}^{(1)}_{(1)})t_{(1)}J^{(1)}b)S^{-1}(Q_J^{-1})$$
$$\times S^{-1}(J^{(2)})S^{-1}(t_{(2)})S(\widetilde{J}^{(1)}_{(2)})\widetilde{J}^{(2)} \quad \text{(by (8.10))}$$
$$= \lambda(S^2(R^{-1})S^2(\widetilde{J}^{-(1)})t_{(1)}J^{(1)}b)S^{-1}(Q_J^{-1})S^{-1}$$
$$\times (J^{(2)})S^{-1}(t_{(2)})S(\widetilde{J}^{-(2)})Q_J \quad \text{(by (8.12))}$$
$$= \lambda(S^2(R^{-1})S^2(\widetilde{J}^{-(1)})S^2(\widetilde{J}^{(1)})S(Q_J^{-1})Q_JJ^{-(1)}J^{(1)}b)$$
$$\times S^{-1}(Q_J^{-1})S^{-1}(J^{(2)})S^{-1}(J^{-(2)})S^{-1}(Q_J)Q_J^{-1}S(\widetilde{J}^{(2)})S(\widetilde{J}^{-(2)})Q_J \quad \text{(by (8.19))}$$
$$= \lambda(S^2(R^{-1})S(Q_J^{-1})Q_Jb)S^{-1}(Q_J^{-1})S^{-1}(Q_J)Q_J^{-1}Q_J$$
$$= \lambda(S^2(R^{-1})S(Q_J^{-1})Q_Jb)$$
$$= \lambda^J(b).$$

Thus, λ^J is a right integral of $(H^J)^*$. The proof is completed. □

Remark 8.4.2 *If H is a finite dimensional unimodular Hopf algebra over a field $\Bbbk$ with a normalized twist J, then*

$$R = \alpha(J^{-(1)})J^{-(2)} = \varepsilon(J^{-(1)})J^{-(2)} = 1.$$

In this case,

$$\lambda^J = \lambda \leftharpoonup S(Q_J^{-1})Q_J,$$

which is exactly the result of [1, *Theorem* 3.4]. *It is interesting to know whether $\lambda(1) \neq 0$ if and only if $\lambda^J(1) \neq 0$ for a finite dimensional Hopf algebra H (see* [1, *Remark* 3.9]). *In the case when* $\mathrm{char}(\Bbbk) = 0$, *then H is semisimple if and only if H is cosemisimple. This implies that $\lambda(1) \neq 0$ if and only if $\lambda^J(1) \neq 0$. In the case when* $\mathrm{char}(\Bbbk) > 0$, *then the equivalence $\lambda(1) \neq 0$ if and only if $\lambda^J(1) \neq 0$ holds when H is unimodular (see* [1, *Corollary* 3.6]).

For any finite dimensional Hopf algebra H, recall from [42] and [72] that the n-th indicator of H is defined by

$$\nu_n(H) := \mathrm{tr}(S \circ P_{n-1}), \quad \text{for } n \in \mathbb{Z}.$$

By Radford's trace formula (8.5), we have

$$\nu_n(H) = \begin{cases} (\lambda \circ S)(\Lambda_{(1)}\Lambda_{(2)} \cdots \Lambda_{(n)}), & n \geqslant 1, \\ \lambda(1)\varepsilon(\Lambda), & n = 0, \\ (\lambda \circ S^2)(\Lambda_{(-n)}\Lambda_{(-n-1)} \cdots \Lambda_{(1)}), & n \leqslant -1. \end{cases} \tag{8.20}$$

The remarkable result due to Kashina, Montgomery and Ng [42, Theorem 2.2] states that the indicator $\nu_n(H)$ of a finite dimensional Hopf algebra H is a monoidal invariant of H-mod for any $n \geqslant 1$. Later this result has been extended to the case $n \leqslant 0$ by Shimizu [72, Theorem 3.10], where the proof relies on the case of $n \geqslant 1$ and the linear recurrence relation between the Sweedler power maps. In the sequel, we will use Theorem 8.4.1 to give a uniform proof of the two cases.

Theorem 8.4.3 *Let H be a finite dimensional Hopf algebra over a field $\Bbbk$. The n-th indicator $\nu_n(H)$ is a monoidal invariant of H-mod for any $n \in \mathbb{Z}$.*

Proof Suppose that $\lambda \in H^*$ is a right integral and $\Lambda \in H$ is a left integral such that $\lambda(\Lambda) = 1$. It follows from (8.20) that

$$\nu_n(H) = (\lambda \circ S)(P_n(\Lambda)), \quad \text{for } n \in \mathbb{Z}.$$

Let H^J be the Hopf algebra twisted by a normalized twist J. The n-th indicator of H^J is

$$\nu_n(H^J) = (\lambda^J \circ S^J)(P_n^J(\Lambda)).$$

We claim that

$$\nu_n(H^J) = \nu_n(H), \quad \text{for } n \in \mathbb{Z}.$$

Indeed,

$$\begin{aligned}
\nu_n(H^J) &= (\lambda^J \circ S^J)(P_n^J(\Lambda)) \\
&= \lambda^J(Q_J^{-1} S(P_n^J(\Lambda))Q_J) \\
&= \lambda(\alpha(J^{(1)})S^2(J^{(2)})S(Q_J^{-1})Q_J Q_J^{-1} S(P_n^J(\Lambda))Q_J) \\
&= \lambda(\alpha(J^{(1)})S^2(J^{(2)})S(Q_J^{-1})S(Q_J^{-1}P_n(\Lambda)\alpha(J^{(-1)}) \\
&\quad \times S(J^{(-2)})Q_J)Q_J) \quad \text{(by Theorem 8.1.4(2))} \\
&= \lambda(S(P_n(\Lambda))S(Q_J^{-1})Q_J) \\
&= \lambda(S(P_n(\Lambda))S(Q_J^{-1})S(J^{(1)})J^{(2)}) \\
&= \lambda(\alpha(J^{(2)}_{(1)})S^2(J^{(2)}_{(2)})S(P_n(\Lambda))S(Q_J^{-1})S(J^{(1)})) \quad \text{(by (8.3))} \\
&= (\lambda \circ S)(J^{(1)}Q_J^{-1}P_n(\Lambda)\alpha(J^{(2)}_{(1)})S(J^{(2)}_{(2)})) \\
&= (\lambda \circ S)(J^{(1)}Q_J^{-1}P_{n-1}(\Lambda_{(1)})\Lambda_{(2)}\alpha(J^{(2)}_{(1)})S(J^{(2)}_{(2)})) \\
&= (\lambda \circ S)(J^{(1)}Q_J^{-1}P_{n-1}(\Lambda_{(1)}J^{(2)})\Lambda_{(2)}) \quad \text{(by (8.2))} \\
&= (\lambda \circ S)(J^{(1)}J^{-(1)}S(J^{-(2)})\Lambda_{(1)}P_{n-1}(J^{(2)}\Lambda_{(2)})) \\
&= (\lambda \circ S)(J^{(1)}J^{-(1)}\Lambda_{(1)}P_{n-1}(J^{(2)}J^{-(2)}\Lambda_{(2)})) \quad \text{(by (8.1))} \\
&= (\lambda \circ S)(\Lambda_{(1)}P_{n-1}(\Lambda_{(2)})) \\
&= (\lambda \circ S)(P_n(\Lambda)) \\
&= \nu_n(H).
\end{aligned}$$

If H' is a finite dimensional Hopf algebra such that H'-mod is monoidally equivalent to H-mod, then there exists a normalized twist J of H such that H' is isomorphic via a map σ to H^J as Hopf algebras. Then

$$\sigma \circ S' \circ P'_n = S^J \circ P_n^J \circ \sigma,$$

where S' and P'_n are the antipode and the n-th Sweedler power map of H' respectively. For any $n \in \mathbb{Z}$, we have

$$\begin{aligned}
\nu_n(H') &= \operatorname{tr}(S' \circ P'_{n-1}) = \operatorname{tr}(\sigma^{-1} \circ S^J \circ P_{n-1}^J \circ \sigma) \\
&= \operatorname{tr}(S^J \circ P_{n-1}^J) = \nu_n(H^J) = \nu_n(H).
\end{aligned}$$

The proof is completed. □

We need to point out that Theorem 8.4.1 can be used to prove the equality

$$\mathrm{tr}(S^n) = \mathrm{tr}((S^J)^n), \quad \text{for } n = \pm 1, \pm 2.$$

However, we are not sure Theorem 8.4.1 can be used to prove the equality for all $n \in \mathbb{Z}$. It is known that if H has the Chevalley property, the equality holds for all $n \in \mathbb{Z}$ (see [60, Theorem 4.3]).

Finally, we will use Theorem 8.4.1 to show that the Killing form of a Hopf algebra is invariant under twisting. Let H be a finite dimensional Hopf algebra with antipode S over a field $\Bbbk$. The (left) adjoint representation of H is the map

$$\mathrm{ad} : H \to \mathrm{End}_{\Bbbk}(H), \qquad a \mapsto \mathrm{ad}a,$$

where the $\Bbbk$-linear map $\mathrm{ad}a : H \longrightarrow H$ is given by

$$(\mathrm{ad}a)(b) = a_{(1)} b S(a_{(2)}), \quad \text{for } b \in H.$$

The *Killing form* of the Hopf algebra H is defined by

$$(a, b) = \mathrm{tr}(\mathrm{ad}a \circ \mathrm{ad}b) = \mathrm{tr}(\mathrm{ad}(ab)), \quad \text{for } a, b \in H.$$

The subspace

$$H^{\perp} = \{a \in H \mid (a, b) = 0 \text{ for all } b \in H\}$$

is called the *Killing radical* of H. This radical is an ideal of H but not a Hopf ideal in general.

We denote the adjoint representation of the twisted Hopf algebra H^J by ad^J and the Killing form of H^J by $(-,-)^J$. Namely,

$$(a, b)^J = \mathrm{tr}(\mathrm{ad}^J(ab)).$$

The following result has been given in [84, Corollary 4.3], we present it here with an alternative proof.

Proposition 8.4.4 *The Killing form of a finite dimensional Hopf algebra H is invariant under twisting. Namely, if H^J is the Hopf algebra twisted by a normalized twist J on a finite dimensional Hopf algebra H, then $(a, b) = (a, b)^J$ for all $a, b \in H$. In particular, $H^{\perp} = (H^J)^{\perp}$.*

Proof For $a, b \in H$, we only need to prove that $(a, 1)^J = (a, 1)$ since $(a, b)^J = (ab, 1)^J$ and $(a, b) = (ab, 1)$. Suppose that $\lambda \in H^*$ is a right integral and $\Lambda \in H$ is a left integral such that $\lambda(\Lambda) = 1$. Recall from Lemma 8.1.3 that

$$\Delta^J(\Lambda) = \Lambda_{\langle 1 \rangle} \otimes \Lambda_{\langle 2 \rangle} = Q_J^{-1} \Lambda_{(1)} \otimes \Lambda_{(2)} S(R) Q_J.$$

By Radford's trace formula (8.5), we have

$$(a, 1)^J = \mathrm{tr}(\mathrm{ad}^J(a))$$

$$
\begin{aligned}
&= \lambda^J(S^J(\Lambda_{\langle 2\rangle})\mathrm{ad}^J(a)(\Lambda_{\langle 1\rangle}))\\
&= \lambda^J(S^J(\Lambda_{(2)}S(R)Q_J)\mathrm{ad}^J(a)(Q_J^{-1}\Lambda_{(1)}))\\
&= \lambda^J(Q_J^{-1}S(\Lambda_{(2)}S(R)Q_J)Q_J a_{\langle 1\rangle}Q_J^{-1}\Lambda_{(1)}S^J(a_{\langle 2\rangle}))\\
&= \lambda^J(Q_J^{-1}S(\Lambda_{(2)}S(R)Q_J)Q_J J^{-(1)}a_{(1)}J^{(1)}Q_J^{-1}\Lambda_{(1)}Q_J^{-1}S(J^{-(2)}a_{(2)}J^{(2)})Q_J)\\
&= \lambda(S^2(R^{-1})S(Q_J^{-1})Q_J Q_J^{-1}S(\Lambda_{(2)}S(R)Q_J)\\
&\quad\times Q_J J^{-(1)}a_{(1)}J^{(1)}Q_J^{-1}\Lambda_{(1)}Q_J^{-1}S(J^{-(2)}a_{(2)}J^{(2)})Q_J)\\
&= \lambda(S(\Lambda_{(2)})Q_J J^{-(1)}a_{(1)}J^{(1)}Q_J^{-1}\Lambda_{(1)}Q_J^{-1}S(J^{-(2)}a_{(2)}J^{(2)})Q_J)\\
&= \lambda(S(\Lambda_{(2)})J^{(1)}Q_J^{-1}Q_J J^{-(1)}a_{(1)}\\
&\quad\times \Lambda_{(1)}Q_J^{-1}S(J^{(2)})S(a_{(2)})S(J^{-(2)})Q_J) \quad \text{(by (8.1))}\\
&= \lambda(S^2(S(J^{-(2)})Q_J \leftharpoonup \alpha)S(\Lambda_{(2)})J^{(1)}J^{-(1)}a_{(1)}\\
&\quad\times \Lambda_{(1)}Q_J^{-1}S(J^{(2)})S(a_{(2)})) \quad \text{(by (8.3))}\\
&= \lambda(S(\Lambda_{(2)})J^{(1)}J^{-(1)}a_{(1)}\Lambda_{(1)}S(J^{-(2)})Q_J Q_J^{-1}S(J^{(2)})S(a_{(2)})) \quad \text{(by (8.2))}\\
&= \lambda(S(\Lambda_{(2)})a_{(1)}\Lambda_{(1)}S(a_{(2)}))\\
&= \mathrm{tr}(\mathrm{ad}(a))\\
&= (a,1).
\end{aligned}
$$

Now $H^\perp = (H^J)^\perp$ follows from $(a,b) = (a,b)^J$ for all $a,b \in H$. □

As a consequence, we obtain a monoidal invariant of H-mod as follows:

Corollary 8.4.5 *Let H be a finite dimensional Hopf algebra over a field $\Bbbk$. Then $\dim_{\Bbbk}(H^\perp)$ is a monoidal invariant of H-mod.*

Example 8.4.6 Consider the dihedral group $D_n = \{1, a, \cdots, a^{n-1}, b, ba, \cdots, ba^{n-1}\}$ with $a^n = 1$, $b^2 = 1$ and $aba = b$. Suppose $2n \neq 0$ in $\Bbbk$. If n is odd, then $(\Bbbk D_n)^\perp = 0$. If n is even, then $(\Bbbk D_n)^\perp$ is the ideal of $\Bbbk D_n$ generated by $a^{\frac{n}{2}} - 1$. This is a Hopf ideal of $\Bbbk D_n$, $\dim_{\Bbbk}((\Bbbk D_n)^\perp) = n$ and the quotient Hopf algebra $\Bbbk D_n/(\Bbbk D_n)^\perp$ is isomorphic to $\Bbbk D_{\frac{n}{2}}$ (see [84, Example 3.7]).

Example 8.4.7 Let $H_{n,d}$ denote a generalized Taft algebra over a field $\Bbbk$ with $nd \neq 0$ in $\Bbbk$. As an algebra, $H_{n,d}$ is generated by g and h subject to the following relations:

$$g^n = 1, \quad h^d = 0, \quad hg = qgh,$$

where d divides n and q is a primitive d-th root of unity. As a Hopf algebra, the comultiplication Δ, counit ε and the antipode S of $H_{n,d}$ are given respectively by

$$\Delta(g) = g \otimes g, \quad \Delta(h) = 1 \otimes h + h \otimes g, \quad \varepsilon(g) = 1, \quad \varepsilon(h) = 0,$$

$$S(g) = g^{-1}, \quad S(h) = -q^{-1}g^{n-1}h.$$

The Hopf algebra $H_{n,d}$ has a $\Bbbk$-basis $\{g^i h^j \mid 0 \leqslant i \leqslant n-1, 0 \leqslant j \leqslant d-1\}$ and $\dim_{\Bbbk}(H_{n,d}) = nd$. If $d = n$, then $H_{n,n}$ is the Taft algebra. By a straightforward computation, the Killing radical of $H_{n,d}$ is the ideal of $H_{n,d}$ generated by $g^d - 1$ and h. The dimension of the Killing radical $H_{n,d}^{\perp}$ is $\dim_{\Bbbk}(H_{n,d}^{\perp}) = (n-1)d$, which is a monoidal invariant of the representation category $H_{n,d}$-mod.

Bibliography

[1] Aljadeff E, Etingof P, Gelaki S, Nikshych D. On twisting of finite-dimensional Hopf algebras. J. Algebra, 2002, 256: 484-501.

[2] Andruskiewitsch N, Angiono I, Iglesias A G, et al. From Hopf algebras to tensor categories. Conformal field theories and tensor categories. Berlin, Heidelberg: Springer, 2014.

[3] Archer L. On certain quotients of the Green rings of dihedral 2-groups. J. Pure Appl. Algebra, 2008, 212: 1888-1897.

[4] Auslander M, Reiten I, Smalφ S. Representation Theory of Artin Algebras. Cambridge Studies in Advanced Mathematics, Vol. 36. Cambridge: Cambridge University Press, 1994.

[5] Bakalov B, Kirillov A A. Lectures on tensor categories and modular functors. Vol. 21, University Series Lectures, AMS, 2001.

[6] Barrett J W, Westbury B W. Invariants of piecewise-linear 3-manifolds. Trans. Amer. Math. Soc., 1996, 348(10): 3997-4022.

[7] Benson D J, Carlson J F. Nilpotent elements in the Green ring. J. Algebra, 1986, 104: 329-350.

[8] Benson D J, Parker R A. The Green ring of a finite group. J. Algebra, 1984, 87: 290-331.

[9] Bhargava M, Zieve M E. Factoring Dickson polynomials over finite fields. Finite Fields Appl., 1999, 5(2): 103-111.

[10] Bruillard P, Ng S H, Rowell E, Wang Z. Rank-finiteness for modular categories. J. Amer. Math. Soc., 2016, 29(3): 857-881.

[11] Caenepeel S, Bogdan Ion, Militaru G. The structure of Frobenius algebras and separable algebras. K-Theory, 2000, 19(4): 365-402.

[12] Carlson J F. The dimensions of periodic modules over modular group algebras. Illinois J. Math., 1979, 23(2): 295-306.

[13] Chen H. The Green ring of Drinfeld double $D(H_4)$. Algebr. Represent. Theory, 2014, 17(5): 1457-1483.

[14] Chen H, Mohammed H, Lin W, Sun H. The projective class rings of a family of pointed Hopf algebras of rank two. Bull. Belg. Math. Soc. Simon Stevin, 2016, 23(5): 693-711.

[15] Chen H, Oystaeyen F V, Zhang Y. The Green rings of Taft algebras. Proc. Amer. Math. Soc., 2014, 142: 765-775.

[16] Chen J, Yang S, Wang D. Grothendieck rings of a class of Hopf algebras of Kac-Paljutkin type. Front. Math. China, 2021, 16(1): 29-47.

[17] Chou W S. The factorization of Dickson polynomials over finite fields. Finite Fields Appl., 1997, 3: 84-96.

[18] Cibils C. A quiver quantum group. Commun. Math. Phys., 1993, 157: 459-477.

[19] Darpö E, Herschend M. On the representation ring of the polynomial algebra over perfect field. Math. Z., 2011, 265: 605-615.

[20] Doi Y. Substructures of bi-Frobenius algebras. J. Algebra, 2002, 256: 568-582.

[21] Doi Y. Bi-Frobenius algebras and group-like algebras. Lecture Notes in Pure and Appl. Math., 2004, 237: 143-156.

[22] Doi Y. Group-like algebras and their representations. Commun. Algebra, 2010, 38(7): 2635-2655.

[23] Doi Y, Takeuchi M. BiFrobenius algebras. Contemp. Math., 2000, 267: 67-98.

[24] Domokos M, Lenagan T H. Representation rings of quantum groups. J. Algebra, 2004, 282: 103-128.

[25] Drinfeld V, Gelaki S, Nikshych D, Ostrik V. On braided fusion categories I. Sel. Math. New Ser., 2010, 16: 1-119.

[26] Erdmann K, Green E L, Snashall N, Taillefer R. Representation theory of the Drinfeld doubles of a family of Hopf algebras. J. Pure Appl. Algebra, 2006, 204(2): 413-454.

[27] Etingof P, Gelaki S. On finite dimensional semisimple and cosemisimple Hopf algebras in positive characteristic. Int. Math. Res. Not. IMRN, 1998, (16): 851-864.

[28] Etingof P, Gelaki S, Nikshych D, Ostrik V. Tensor Categories. Mathematical Surveys and Monographs. Vol. 205. Providence: AMS, 2015.

[29] Etingof P, Nikshych D, Ostrik V. An analogue of Radford's S^4 formula for finite tensor categories. Int. Math. Res. Not. IMRN, 2004, 54: 2915-2933.

[30] Etingof P, Nikshych D, Ostrik V. On fusion categories. Ann. Math., 2005, 162: 581-642.

[31] Feigin B L, Gainutdinov A M, Semikhatov A M, Tipunin I Y. Modular group representations and fusion in logarithmic conformal field theories and in the quantum group center. Commun. Math. Phys., 2006, 265(1): 47-93.

[32] Geck M, Pfeiffer G. Characters of Finite Coxeter Groups and Iwahori-Hecke Algebras. New York: Oxford University Press, 2000.

[33] Green J A. The modular representation algebra of a finite group III. J. Math., 1962, 6(4): 607-619.

[34] Green J A. A transfer theorem for mudular representations. J. Algebra, 1964, 1: 73-84.

[35] Green E L, Marcos E N, Solberg Φ. Representations and almost split sequences for Hopf algebras. Representation Theory of Algebras: Seventh International Conference on Representations of Algebras, Cocoyoc, Mexico, Vol. 18. Providence: AMS, 1994.

[36] Gunnlaugsdöttir E. Monoidal structure of the category of u_q^+-modules. Linear Algebra Appl., 2003, 365: 183-199.

[37] Haim M. Group-like algebras and Hadamard matrices. J. Algebra, 2007, 308: 215-235.

[38] Happel D. Triangulated Categories in the Representation of Finite Dimensional Algebras. Cambridge: Cambridge University Press, 1988.

[39] Higman D G. On orders in separable algebras. Canad. J. Math., 1955, 7: 509-515.

[40] Huang H, Oystaeyen F V, Yang Y, Zhang Y. The Green rings of pointed tensor categories of finite type. J. Pure Appl. Algebra, 2014, 218: 333-342.

[41] Huang H, Yang Y. The Green rings of minimal Hopf quivers. Proc. Edinb. Math. Soc., 2016, 59(1): 107-141.

[42] Kashina Y, Montgomery S, Ng S H. On the trace of the antipode and higher indicators. Israel J. Math., 2012, 188(1): 57-89.

[43] Kashina Y, Sommerhäuser Y, Zhu Y. On higher Frobenius-Schur indicators. Mem. Amer. Math. Soc., 2006, 181(855): viii+65.

[44] Kassel C. Quantum Groups. New York: Springer-Verlag, 1995.

[45] Landers R, Montgomery S, Schauenburg P. Hopf powers and orders for some bismash products. J. Pure Appl. Algebra, 2006, 205(1): 156-188.

[46] Larson R G, Radford D E. Finite dimensional cosemisimple Hopf algebras in characteristic 0 are semisimple. J. Algebra, 1988, 117: 267-289.

[47] Li Y, Hu N. The Green rings of the 2-rank Taft algebra and its two relatives twisted. J. Algebra, 2014, 410: 1-35.

[48] Li K, Liu G. On the antipode of Hopf algebras with the dual Chevalley property. J. Pure Appl. Algebra, 2022, 226(3): 106871.

[49] Li L, Zhang Y. The Green rings of the generalized Taft Hopf algebras. Contemp. Math., 2013, 585: 275-288.

[50] Linchenko V, Montgomery S. A Frobenius-Schur theorem for Hopf algebras. Algebr. Represent. Theory, 2000, 3(4): 347-355.

[51] Lorenz M. Representations of finite-dimensional Hopf algebras. J. Algebra, 1997, 188: 476-505.

[52] Lorenz M. Some applications of Frobenius algebras to Hopf algebras. Contemp. Math., 2011, 537: 269-289.

[53] Masuoka A. Semisimple Hopf algebras of dimension 6, 8. Israel J. Math., 1995, 92: 361-373.

[54] McDonald B R. Finite Rings with Identity. Vol. 28. Marcel Dekker Incorporated, 1974.

[55] Molnar R K. Semi-direct products of Hopf algebras. J. Algebra, 1977, 47(1): 29-51.

[56] Montgomery S. Hopf Algebras and Their Actions on Rings. CBMS Series in Math, vol.82. Providence: Amer. Math. Soc., 1993.

[57] Montgomery S, Witherspoon S J. Irreducible representations of crossed products. J. Pure Appl. Algebra, 1998, 129(3): 315-326.

[58] Müger M. From subfactors to categories and topology. II. The quantum double of tensor categories and subfactors. J. Pure Appl. Algebra, 2003, 180(1-2): 159-219.

[59] Natale S. Hopf algebras of dimension 12. Algebr. Represent. Theory, 2002, 5(5): 445-455.

[60] Negron C, Ng S H. Gauge invariants from the powers of antipodes. Pacific J. Math., 2017, 291(2): 439-460.

[61] Ng S H, Schauenburg P. Frobenius-Schur indicators and exponents of spherical categories. Adv. Math., 2007, 211(1): 34-71.

[62] Ng S H, Schauenburg P. Higher Frobenius-Schur indicators for pivotal categories. Hopf Algebras and Generalizations. Contemp. Math. 2007, 441: 63-90.

[63] Ng S H, Schauenburg P. Central invariants and higher indicators for semisimple quasi-Hopf algebras. Trans. Amer. Math. Soc., 2008, 360(4): 1839-1860.

[64] Ng S H, Schauenburg P. Congruence subgroups and generalized Frobenius-Schur indicators. Comm. Math. Phys., 2011, 300(1): 1-46.

[65] Nichols W D, Richmond M B. The Grothendieck algebra of a Hopf algebra I. Commun. Algebra, 1998, 26(4): 1081-1095.

[66] Ostrik V. On formal codegrees of fusion categories. Math. Res. Lett., 2009, 16(5): 895-901.

[67] Ostrik V. Pivotal fusion categories of rank 3. Mosc. Math. J., 2015, 15(2): 373-396.

[68] Ostrik V. On symmetric fusion categories in positive characteristic. Sel. Math. New Ser., 2020, 26(3): 1-19.

[69] Radford D E. The order of the antipode of a finite dimensional Hopf algebra is finite. Amer. J. Math., 1976, 98: 333-355.

[70] Radford D E. The trace function and Hopf algebras. J. Algebra, 1994, 163: 583-622.

[71] Santos W F, Haim M. Radford's formula for bi-Frobenius algebras and applications. Commun. Algebra, 2008, 36(4): 1301-1310.

[72] Shimizu K. On indicators of Hopf algebras. Israel J. Math., 2015, 207(1): 155-201.

[73] Shimizu K. The monoidal center and the character algebra. J. Pure Appl. Algebra, 2017, 221(9): 2338-2371.

[74] Siehler J. Near-group categories. Alg. Geom. Topol., 2003, 3(2): 719-775.

[75] Skowroński A, Yamagata K. Frobenius Algebras. Zürich: European Mathematical Society, 2011.

[76] Sommerhäuser Y. On Kaplansky's fifth conjecture. J. Algebra, 1998, 204: 202-224.

[77] Su D, Yang S. Green rings of weak Hopf algebras based on generalized Taft algebras. Period. Math. Hung., 2018, 76: 229-242.

[78] Sun H, Chen H. Green ring of the category of weight modules over the Hopf-Ore extensions of group algebras. Commun. Algebra, 2019, 47(11): 4441-4461.

[79] Sun H, Mohammed H, Lin W, Chen H. Green rings of Drinfeld doubles of Taft algebras. Commun. Algebra, 2020, 48(9): 3933-3947.

[80] Sweedler M E. Hopf Algebras. New York: Benjamin, 1969.

[81] Tambara D, Yamagami S. Tensor categories with fusion rules of self-duality for finite abelian groups. J. Algebra, 1998, 209: 692-707.

[82] Wakui M. Various structures associated to the representation categories of eight-dimensional nonsemisimple Hopf algebras. Algebr. Represent. Theory, 2004, 7: 491-515.

[83] Wang Y, Chen X. Construct non-graded bi-Frobenius algebras via quivers. Sci. China Ser. A-math., 2007, 50(3): 450-456.

[84] Wang Z, Chen H, Li L. The Killing form of a Hopf algebra and its radical. Arab. J. Sci. Eng. Sect., 2008, 33(2C): 553-559.

[85] Wang Z, Li L. On realization of fusion rings from generalized Cartan matrices. Acta. Math. Sin.-English Ser., 2017, 33(3): 362-376.

[86] Wang Z, Li L, Zhang Y. Green rings of pointed rank one Hopf algebras of nilpotent type. Algebr. Represent. Theory, 2014, 17(6): 1901-1924.

[87] Wang Z, Li L, Zhang Y. Green rings of pointed rank one Hopf algebras of non-nilpotent type. J. Algebra, 2016, 449: 108-137.

[88] Wang Z, Li L, Zhang Y. A criterion for the Jacobson semisimplicity of the Green ring of a finite tensor category. Glasg. Math. J., 2018, 60(1): 253-272.

[89] Wang Z, Li L, Zhang Y. Bilinear forms on Green rings of finite dimensional Hopf algebras. Algebr. Represent. Theory, 2019, 22(6): 1569-1598.

[90] Wang Z, Liu G, Li L. The Casimir number and the determinant of a fusion category. Glasg. Math. J., 2021, 63(2): 438-450.

[91] Wang Z, Liu G, Li L. Invariants from the Sweedler power maps on integrals. J. Algebra, 2022, 606: 590-612.

[92] Wang Y, Zhang P. Construct bi-Frobenius algebras via quivers. Tsukuba J. Math., 2004, 28(1): 215-227.

[93] Witherspoon S J. The representation ring of the quantum double of a finite group. J. Algebra, 1996, 179: 305-329.

[94] Witherspoon S J. The representation ring and the centre of a Hopf algebra. Canad. J. Math.,1999, 51(4): 881-896.

[95] Yang S. Finite dimensional representations of u-Hopf algebras. Commun. Algebra, 2001, 29(12): 5359-5370.

[96] Yang R, Yang S. The Grothendieck rings of Wu-Liu-Ding algebras and their Casimir numbers (II). Commun. Algebra, 2021, 49(5): 2041-2073.

[97] Yang R, Yang S. The Grothendieck rings of Wu-Liu-Ding algebras and their Casimir numbers (I). J. Algebra Appl., 2022, 21(9): 2250178.

[98] Zemanek J. Nilpotent elements in representation rings. J. Algebra, 1971, 19: 453-469.

[99] Zhu Y. Hopf algebras of prime dimension. Int. Math. Res. Not. IMRN, 1994, 1: 53-59.

Index

编 后 记

《博士后文库》是汇集自然科学领域博士后研究人员优秀学术成果的系列丛书.《博士后文库》致力于打造专属于博士后学术创新的旗舰品牌，营造博士后百花齐放的学术氛围，提升博士后优秀成果的学术和社会影响力.

《博士后文库》出版资助工作开展以来，得到了全国博士后管委会办公室、中国博士后科学基金会、中国科学院、科学出版社等有关单位领导的大力支持，众多热心博士后事业的专家学者给予积极的建议，工作人员做了大量艰苦细致的工作. 在此，我们一并表示感谢！

《博士后文库》编委会